Confident Mathematics Teaching 5–13:
INSET in the classroom

Confident Mathematics Teaching 5 to 13: INSET in the classroom

Edith Biggs

NFER-Nelson

Published by The NFER-Nelson Publishing Company Ltd.,
Darville House, 2 Oxford Road East,
Windsor, Berks, SL4 1DF

First published 1983
© Edith Biggs, 1983
ISBN 0 7005 0581 4
Code 8148 021

Photoset in Sabon by Rowland Phototypesetting Ltd
Bury St Edmunds, Suffolk
Printed in Great Britain

Distributed in the USA by Humanities Press Inc.,
Atlantic Highlands, New Jersey 07716 USA.

Contents

List of tables

Foreword

This book gives an account of the struggles, set-backs and successes of all those concerned with an in-service project in mathematics within 14 schools in an outer metropolitan borough. There were periods of exhilaration as well as of depression; gradually, almost imperceptibly, the exhilaration gained ground over the depression. But there were failures as well as successes – not least because of the unexpectedly high staff turnover during the period April 1976 to August 1979. I wish to take this opportunity of thanking all those who took part with me in the project – children, teachers, heads, advisers and lecturers – for their co-operation at all stages, for the frankness with which they expressed their views and for all the hard work they put in. Many of them became my friends. It was a valuable learning experience for us all.

When I began this project I knew little about non-statistical methods of research in the field of education. Much of what I subsequently learned was under the guidance of Miss Helen Simons at the London Institute of Education. I wish to thank her for many stimulating sessions and for introducing me to several researchers already working in the field of case study. I also wish to thank Professor Denis Lawton, my supervisor, who gave me unfailingly sound advice and sustained encouragement. He helped me to structure my findings and convinced me of the validity of these ways of working. Finally, I thank Miss Kay Burton for helping me to say what I wanted to say; I appreciate her expertise.

CHAPTER 1

Setting the scene

The state of mathematics education

During the past ten years teachers have been made uncertain about the effectiveness of their teaching methods by public criticism of mathematics teaching. This criticism focused attention on falling standards and recommended a move 'back to basics'. At the same time teachers felt that their teaching of basics was undermined by the availability of cheap electronic calculators. They asked such questions as: Is it still necessary for children to learn their tables? Do they need to understand the processes of the four operations? Should they be given practice in four-figure calculations? The accusations about falling standards were said to come mainly from employers criticizing the performance of 16-year-olds in traditional arithmetic tests. 'Is it fair,' teachers asked, 'to use tests of this kind when the scope of mathematics has been broadened? Have standards of achievement in mathematics really changed as much as employers claim?'

Because of the fears that standards had fallen the Department of Education and Science (DES) took the first step towards monitoring national standards of attainment in mathematics. In the early 1970s a special group was set up to prepare 'Tests of Attainment in Mathematics' (TAMS) (Summer, 1975). An item bank of tests was prepared for pupils at the ages of 11 and 16. The tests were to be administered annually to a random selection of pupils in these age groups in a random selection of schools in England and Wales. The first tests required only written answers, but subsequently a new type of test was devised: practical assessments for individual children. These tests were planned so that each practical activity could also be given as a paper and pencil test. All the children involved were given both written and practical versions with an interval of at least three weeks between the two types. The report on this work was of special interest:

> More pupils scored high marks on the practical test than on the written test . . . The differences between the two methods of testing show quite clearly that one is not equivalent to the other and that neither should replace the other . . . Where the practical version was attempted first . . . the score on the written test is considerably improved . . . It appears that there was a considerable learning effect when the practical version preceded the written version.

These findings were encouraging for those of us who were advocating the preparation of structured practical activities to help children to learn mathematics.

The results of the TAMS experiment were used as a basis for the material devised by the Assessment of Performance Unit (APU), set up in 1977 to assess national standards of achievement across the curriculum. A description of the work of the APU included the statements:

> Some [assessments] will test skills, some concepts and some applications. The major emphases of the APU in mathematics are on problem solving and applications. The questions are asked in a fresh context with a real-life flavour.

It will be interesting to observe what happens to the mathematics curriculum as the content of the APU practical assessments becomes more widely known.

Reports by Her Majesty's Inspectors (HMI) during the past three years have borne out my own assessment of the teaching of mathematics up to the age of 13. A recent major report on primary education (GB. DES, 1978) was based on the direct observation of children's work by HMIs, all of whom were experienced in this field. The report included the following statements:

> Efforts made to teach children to calculate are not rewarded by high scores in the examples concerned with the handling of everyday situations. Learning to operate with numbers may need to be more closely linked with learning to use them in a variety of

situations than is now common . . . The teaching of skills in isolation, whether in language or mathematics, does not produce the best results.

A second document (GB. DES, 1979) was written by HMIs with a special knowledge of mathematics. This discussion document refers to the major attacks:

> Recently, there has been a measure of reaction against innovation, and there have been protests from some critics of the educational system that certain traditional standards are not being adequately maintained. In a minority of primary schools, mathematics is taught much as it was 20 years ago. In others a more practical and informal approach has been adopted and new content has been introduced . . .

The report concludes:

> There is room for improvement in the teaching of mathematics, but practices vary so much from school to school that there are no universal remedies.

Nevertheless, I hoped that the research I was undertaking would find some remedies for the problems of the schools participating in the project.

Despite the efforts to monitor standards in mathematics, public anxiety continued. In consequence an inquiry was set up in 1979 by the Secretary of State for Education and Science (SSES) into the teaching of mathematics in primary and secondary schools in England and Wales. The results of this three-year inquiry (Cockcroft Report, 1982) should reassure teachers. For example:

> The overall picture which emerged is much more encouraging than earlier complaints had led us to expect. We have found little real dissatisfaction amongst employers with the mathematical capabilities of those whom they recruit from schools except in respect of entrance to the retail trade and to engineering apprenticeships.

There was also support in the report for the broadening of the curriculum (which is the reverse of the 'back to basics' recommendation):

We believe that this broadening of the curriculum has had a beneficial effect both in improving children's attitudes to mathematics and also in laying the foundations for a better understanding.

(But the report warns that 'work of this kind needs to be carefully structured and followed up by the teacher'.) The report continues:

The primary mathematics curriculum should enrich children's aesthetic and linguistic experience, provide them with the means of exploring their environment and develop their powers of logical thought, in addition to equipping them with the numerical skills which will be a powerful tool for later work and study.

There is support, too, for the use of electronic calculators:

The use of calculators has not produced any adverse effect on basic computational ability . . . The availability of a calculator in no way reduces the need for mathematical understanding on the part of the person using it . . . There can be little doubt of the motivating effect which calculators have for very many children even at an early age.

Despite the reassurance contained in this report, the committee made it clear that there is considerable room for improvement in the teaching of mathematics in schools, particularly by means of in-service education.

The need for in-service education in mathematics

During the past 20 years some changes have been made in mathematics teaching in two major respects: content and method. Reference has already been made to the broadening of the mathematics curriculum at the primary phase, which now frequently includes shapes, logical thinking and some practical work in all the measures. The first changes in teaching methods were made by teachers of the youngest children, in other aspects of the curriculum than mathematics. Efforts were made by some teachers to help children to be active in their learning rather than to be passive recipients of information. In mathematics this means that children are helped to develop their own solutions to problems and their own methods of calculation, rather than being shown one method by the teacher and practising that alone. These teachers structure situations to help children to acquire concepts; they recognize the importance of questioning children to help their learning and of stimulating discussion among them. Overall, these teachers try to show mathematics as an attractive and creative subject rather than as a subject confined to the acquisition of calculating techniques, most of which are learned by rote.

However, the majority of teachers still find mathematics a difficult subject to teach and tend to rely on letting children work through textbooks or workcards without any practical work or discussion. The basic problem is that few, if any, teachers have learned mathematics through investigation themselves. Most adults learned the subject by rote. Many are aware neither of the situations which help children to acquire concepts nor of the language patterns which should accompany the situations. There is a great need for in-service education which would help teachers to learn mathematics in an active way so that they appreciate the advantages of exchanging ideas with their peers and comparing different methods. By this means they would become convinced, at first hand, of the value to children of learning mathematics through investigation and discussion with their peers.

My own experience within in-service education in mathematics

When I retired towards the end of 1974 I had already spent a good deal of the previous 20 years working in the field of in-service education in mathematics (covering the age range 5 to 13) as part of my responsibilities as HMI. The in-service education with which I was concerned consisted almost entirely of workshop sessions. The teachers were organized in groups and worked through a variety of activities involving those broader aspects of mathematics appropriate for primary children. They also planned sequences of activities for the classroom which it was hoped they would try with the children they taught. A follow-up workshop was always planned, to which the teachers were encouraged to bring a sequence of work they had tried with the children. But it soon became apparent that the effects of the workshops on teachers were not lasting, with a few notable exceptions, mainly of those who had a special interest in mathematics and who were prepared to sustain the impetus I had been able to provide. This run-down occurred despite the fact that most teachers attending the workshops came of their own choice, and were presumably willing to make changes.

The content of the working sessions gradually evolved to include more work on numbers (e.g. place value, the four operations, fractions), as I came to realize how anxious teachers were about the teaching of these aspects of mathematics. Furthermore, in an attempt to make the workshops more effective I experimented with their duration and the interval between the initial and follow-up sessions. The workshops were not confined to the UK, but the results were almost always the same. There was an enthusiastic response initially, and much interesting work was carried out with children during the first six months or so. But gradually the impetus was lost, probably as teachers came to the end of their mathematical resources.

Many of these courses led to the setting up of a team of teachers (usually led by an LEA mathematics adviser) to prepare mathematics guidelines for all teachers of primary children in the area. This involved the team in several meetings and classroom trials, often over a period of two years. The teachers in the team all benefitted from the experience, but when the guidelines were distributed to other schools the teachers who had not been in the team found the guidelines difficult to implement. It seemed that teachers had to be involved in the preparation of the guidelines themselves if they were to be able to put these into practice.

What was the cause of the loss of impetus generated by the workshops? Were we expecting too much of teachers – to broaden the content of mathematics and at the same time to change their teaching style in order to provide children with structured activities and opportunities for discussion? The workshops helped to inform teachers about the new content and a different way of working, and helped them to plan specimen activities, but could not give them the classroom help they needed when they did not know what question to ask next, let alone how to continue a topic. Perhaps the teachers needed help in their classrooms until they gained confidence.

I decided that I would experiment with reinforcing working sessions by providing classroom support for individual teachers. The working sessions would serve several purposes: helping teachers to enjoy mathematics; increasing their confidence in their own ability to learn the subject with understanding; providing specimen sequences of activities for a few key concepts as examples of the kind of planning teachers needed to do for all aspects of mathematics.

In the past I had worked mainly with teachers who already wanted to change their teaching of mathematics. Now I was determined to offer help to every teacher in project schools. My aim would be to ensure that as many teachers as possible had the opportunity to change their teaching style, with my help in the classroom as well as at working sessions. At the same time, I decided to compare the effects on the teaching of mathematics of on-site working sessions which would be attended by the head and all the teachers, with working sessions at a teachers' centre for a key team of three or four teachers from each school. I calculated that I should be able to work extensively with the teachers of six infant and six junior schools, and peripherally with the first year teachers of two of the associated high schools.

The choice of Local Education Authority and arrangements made with them

The area (an outer metropolitan borough) was chosen for a number of reasons.

1. It was easily accessible from my home.
2. Although I had formerly run a few workshops in the area I did not know the schools.
3. The area had a good mixture of socio-economic classes.
4. The borough was developing a comprehensive advisory service in education.

There was one adviser for mathematics; since she was relatively senior there were additional calls on her time. Furthermore, because of the prolonged illness of the chief adviser at the beginning of the project, the mathematics adviser had to take over many of her responsibilities and was unable to play the active part in the project which she had intended. There was also a part-time advisory teacher for mathematics who divided his time between working with individual teachers at first and middle schools and conducting mathematics workshops at the teachers' centre. (For the remainder of his time he was a mathematics lecturer at the local college of education.) I had worked with both previously and was sure of their co-operation.

From the beginning, the proposal of a mathematics project was warmly welcomed by the director of education. He hoped that all the advisers would take part and also suggested that the two full-time mathematics lecturers at the local college of education should be included in the team of helpers, since they were giving valuable assistance at the teachers' centre. (They had already agreed to take part in the project.)

The chief adviser and the mathematics adviser selected the schools which would be invited to participate. Three first and three middle schools in each of two contrasting areas were selected: one downtown area (area 2) and a more suburban area where the fathers were skilled and semi-skilled (area 1). (To prevent identification the first schools have been called Frame, Fleet, Foster, Flanders, Fowler and Finlay; and the middle schools: Melia, Meakins, Missingham, Movehall, Makewell and Measures. Area 1: Frame, Fleet, Foster and Melia, Meakins, Missingham; area 2: Flanders, Fowler, Finlay and Movehall, Makewell, Measures.) At this meeting I learned of two developments in the borough which would undoubtedly affect the outcomes of the project. In September, 1974, the existing infant, junior and secondary schools were reorganized as first schools (ages 5 to 8), middle schools (ages 8 to 12) and high schools (ages 12 to 18). This reorganization had caused considerable movement of staff. Some new heads were appointed, most of whom were in their first headships. Some teachers from the younger classes of junior schools were transferred to the first schools. Many of these teachers had always taught their class as a whole. Since the reorganization was so recent, the new heads and transferred teachers had not had sufficient time to settle in their new posts before the start of the project. Moreover, some schools were waiting to transfer to new buildings and were sharing premises in the meantime.

The second development might influence the project even more. All first and middle schools in the borough were empowered by the Local Education Authority to appoint a mathematics co-ordinator (a scale post). It remained to be seen how heads would set about making such appointments, and what the LEA would do to train the co-ordinators. I welcomed this plan because the co-ordinator in each school would be the natural point of contact with the project. I hoped that a co-ordinator appointed by the head of a school would automatically acquire status in the school.

Despite the willingness of the director of education for schools from his borough to take part in the project, certain restrictions had to be imposed. Although I had planned five one-day working sessions, because of understaffing in the school during 1975–6, no school could release three teachers simultaneously for a whole day at a time. Ultimately it was agreed that teams of key teachers should be released for five afternoons (later extended to seven) at weekly intervals. The teachers at schools with the on-site pattern of working sessions had to work through the morning and for one hour in the afternoon before they could be released. This modification of the programme was a distinct disadvantage because all the teachers would arrive tired after a morning's work. The final programme of working sessions and support visits was eventually agreed by the mathematics advisers.

Outcome of a conference for the advisers

With the mathematics advisers I organized a conference lasting two and a half days for all the members of the advisory team. One major purpose was to inform them of the aims of the project and to give them experience, at first hand, of some of the activities planned for the first input. The part the advisers would be expected to play (giving encouragement to teachers who were trying to make changes in their teaching of mathematics) was also outlined. Much interest was shown and discussion continued after the conference. But it soon became apparent that a number of the advisers were dismayed at the prospect of encouraging teachers in innovations to be made in the teaching of mathematics, even in the schools in which they were responsible. Finally, in consequence of· the good offices of the mathematics adviser, a team of six advisers volunteered their help with the project. The team included two advisers with major re-sponsibilities in first and middle schools.

At this stage it was decided that the advisers (and the two mathematics lecturers from the local college of education) should give assistance in two ways: observation visits and support visits to all the project schools. The observation visits were to be made before the first input. Their main purpose was to obtain an overall picture of the teaching of mathematics in each school, concentrating on the co-ordinator and key teachers.

The advisers had originally set aside one day for observation visits, but since they had agreed (in the interests of objectivity) to work in schools they did not know, they subsequently realized that they would have to pay more than one visit. But the increasing demands made on the advisers' time, partly caused by a high staff turnover which necessitated interviewing for new appointments as well as for promotion, and partly by in-service requirements related to the reorganization of schools, made further visits impossible until after the completion of the first input of in-service. By then some changes had already taken place.

The contribution of recent research into teaching styles

As yet, research into the relative merits of class and group teaching has been inconclusive, as the contrasting results from two research projects show (Bennett, 1976, and Horwitz, 1976). (The size of Bennett's sample was nearly twice that of Horwitz's.) Both Bennett and Horwitz used tests of reading skills, but Bennett included connected writing and computation, whereas Horwitz based his results also on IQ, tests of creativity and the children's attitudes to school and learning. Bennett's investigation favoured traditional teaching methods, while Horwitz's study favoured informal teaching. More recently, Bennett has used different statistical techniques to reinterpret his data. He has now found that the difference between the results for the two teaching styles is not statistically significant. The debate is still open. (NFER is currently carrying out a large-scale investigation, to be completed in 1984, into the relative merits of class and group teaching. Perhaps the results of this study will be more definitive.)

Group work is normally associated with informal teaching (also referred to as open or progressive teaching, as opposed to closed and traditional styles). Informal teaching of this kind is often associated with 'discovery learning'. (I prefer to use the description 'learning by investigation' because this covers pencil and paper investigations as well as problems requiring the use of material.) In a study by an educational psychologist (Ausubel *et al.*, 1968) usually critical of discovery learning by adolescents it was stated that:

In the early, unsophisticated stages of learning any abstract subject matter, particularly prior to adolescence, the discovery method is extremely helpful . . . The discovery method also has obvious uses in evaluating learning outcomes and in teaching problem-solving techniques and appreciating scientific method . . . There is no better way of developing effective skills . . . toward the possibility of solving problems of one's own.

Problem-solving is obviously essential in mathematics. A recommendation of discovery methods as the best route to problem-solving, from someone who is far from being a devotee, must be given considerable weight when teaching methods for primary mathematics are being chosen.

Recently, Galton, Simon and Croll (1980) made a survey of teaching skills in primary schools. They found that individual work seemed to be gaining ground at the expense of group work. They concluded that 'The overall pattern is still, however, fairly traditional.' They emphasized the deprivation of pupils who rarely have an opportunity for discussion. They also criticized the colleges of education for failing to prepare students to evaluate the quality of pupil interaction. They wrote:

Those responsible for training teachers seem to spend little time in teaching their students how to evaluate the quality of pupil–pupil interactions taking place or even to increase the number above the dismally low proportion at present occurring.

The research methods used for the project

I decided to use methods of research new to me: action research and case studies. The use of statistical methods in educational research has been criticized for a number of years, on the grounds that it places too much emphasis on the collection of numerical data obtained under controlled conditions and too little on classroom observation. In statistical research the schools are not considered as individual wholes with complex interrelated problems; no account is taken of circumstances which were not covered in the original research design. In this type of research the researcher has to function as evaluator of what has been done by others. The aim is to ascertain the truth of the situation and not to influence the action of others. But the purpose of my research was to effect a change in the situation by influencing the action of teachers; my major aim was to improve the teaching of mathematics in 12 first and middle schools. Such a method is now called 'action research'. Others have described action research in the following ways:

'Action research is concerned with the everyday practical problems experienced by teachers.' (Elliott, 1978)

'The objective is to get something done.' (Halsey, 1972)

'It is usually collaborative (researchers and practitioners work together on a project). It is self-evaluative – modifications are continuously evaluated within the on-going situation, the ultimate objective being to improve practice in some way or other.' (Cohen and Manion, 1980)

In the first instance the role of change-agent would be mine alone, but I hoped that eventually the co-ordinators, supported by the heads, would become the main change-agents in the classroom. I planned to ask for the assistance of heads and teachers throughout the duration of the project, particularly in appraising the working sessions and support visits. (In the event, not only the advisers but also the college lecturers were too hard-pressed by increased administrative demands to be able to give the time they originally promised to the project. The co-operation of the heads and the teachers, particularly the frankness of their appraisal of the progress of the project from time to time, was invaluable. Without this help the project could not have been completed.)

In case study, information is collected by means of working alongside the teachers and observing them in their classrooms, and by interviews and questionnaires. Observation, therefore, was to be an important process in building up case studies. In deciding on the characteristics I should look for (and ask the advisers to look for) I read widely. I knew that much research on observation had been completed in the USA, but since a large part had been geared to class teaching, the observation schedules so far published were not appropriate for less formal teaching.

Harlen's research techniques (1977) in the UK were more relevant to my problems. She described what she meant by observation:

Observation doesn't just mean watching or looking and it isn't necessarily time-consuming, because it can be carried out as part of normal interaction with the children.

For Harlen, observation was a matter of listening, and attending to what was said, discussing with individuals their work or ideas, noticing how they behaved with other children.

There would be three characteristics which it would be specially important for the advisers and myself to observe: the number and scope of any activities provided for the children; the extent of understanding the teachers had of the purpose of each activity; the amount of discussion in which the children were allowed to participate. I hoped to continue my own observations at the first support visits, before the teachers had begun to make maximum use of these visits.

Case studies for each school would be constructed from the extensive information collected in this way. Nothing was to be excluded from consideration, even things which had not been anticipated at the outset. The important factors would emerge from a scrutiny of all the case studies taken together.

Their peculiar strength lies in their attention to the subtlety and complexity of the case in its own right. (Adelman, Jenkins and Kemmis, 1976)

It is the emergence of recurring factors in the case studies which forms the basis for this book.

My own role as change-agent was, to some extent, conditioned by two criticisms that had been made before the research began. It was suggested that I had two unfair advantages which might invalidate the results of my research: my former wide experience of in-service education in mathematics; my former status as HMI.

The first accusation implied that, even if I succeeded in changing teachers' attitudes and the methods they used in teaching mathematics, the research could not be replicated by others. The second implied that teachers could be intimidated and influenced against their will and better judgement. To counter the second accusation I decided to work in schools where I was not known as HMI, and to adopt as low a key as possible, particularly since I was anxious to affect as many teachers as I could. It occurred to me that my loss of status might have an opposite effect on heads and teachers: they could be reluctant to make the effort required for changes to be made. My supposition was supported when the head of a school in one area reported on the animosity which had been displayed by heads of project schools towards releasing a team of three teachers for one afternoon a week for five weeks. 'Who the hell does she think she is?' one head asked. 'How do we know whether or not she is out of date and past it?' another continued.

The possibility of refuting the first criticism arose in an unexpected way. I discovered during the first input that the mathematics advisory teacher was working on a regular basis with the teachers in 12 first and middle schools in the borough. He gave individual teachers in these schools support in their classrooms and worked with all the teachers at a school at the end of the school day. This was clearly an experiment parallel to mine. I hoped that this advisory teacher's work would show that my research could be replicated by others. At that time he had not had much experience in structuring practical activities, either for students or teachers, as a basis for learning. This made his experiment even more valuable.

Preliminary visits to project schools

The first stage of the project was the preliminary visits to the 14 project schools. These were of four types.

1. Introductory visits to the head and the teachers to explain the aims of the project.
2. Interviews with selected children to assist us in the planning of the working sessions.
3. Observation visits to key teams of teachers to determine the range of teaching styles used.
4. Interviews with selected teachers to form the basis for a study of the attitudes to mathematics of all the teachers in project schools.

This attitude study forms the content of Chapters 2 and 3.

CHAPTER 2

Teachers' attitudes to mathematics

Introduction

While considering the design of the project and the problems I might face in helping the teachers to make changes in their teaching of mathematics, I realized that I ought first to ascertain their attitudes to the subject. I accepted that a large majority might not be willing to say that they lacked confidence when teaching the subject. Nevertheless, if they volunteered that they had disliked mathematics at school or college, they might be willing to tell me when the dislike first began and the circumstances which caused it. It would also be useful to discover whether the teachers remembered their professional training in mathematics and whether they thought it had been adequate.

I therefore decided to investigate teachers' attitudes to mathematics while (a) at school, (b) at college and (c) while teaching it. I planned to do this in two ways.

1. By asking every head and teacher, before the project began, to assess their attitudes to mathematics during these three periods on a five-point scale from A, positive, to E, negative. This scale was chosen to preclude too many neutral assessments.

2. By questionnaire, based on statements made by teachers and heads during interviews. I had read an account of a study carried out by Tuppen (1965) into the attitudes of teachers in junior schools to streaming:

A questionnaire was constructed containing statements which had actually been made by the teachers in the interviews. In the questionnaire . . . each teacher was asked to indicate his degree of agreement or disagreement with each statement, using a five-point scale with a choice of responses ranging from 'strongly agree' to 'strongly disagree'. It is important to notice that the questionnaire was based upon the concepts of practising teachers and was phrased in their language.

I decided that I would use this method for constructing a questionnaire for all the heads and teachers in project schools to complete. I then hoped to be able to compare the attitudes expressed by the teachers in the questionnaires and in the assessments.

The assessments

The heads and teachers were asked to give their own assessments of their attitudes to mathematics without previous discussion with their colleagues. Table 2.1 shows the total assessments made in each category from A to E by the heads and teachers, school by school. In Table 2.1 the category which includes the head's assessment is ringed. As might be expected, the first impression the table gives is that attitudes to teaching mathematics are very different from attitudes to the subject at school and at college. Few teachers at either first or middle schools were willing to confess that they were not confident when teaching mathematics: under 10 per cent expressed negative attitudes; 17 per cent of teachers at first schools and 27 per cent of those at middle schools assessed their attitudes to teaching mathematics as neutral.

Table 2.1 also shows that only one of the six first school heads had a positive attitude to mathematics at school, while two had negative attitudes and three neutral. The corresponding assessments of their attitudes to mathematics at college were two positive, two negative and two neutral. In the six middle schools three heads gave positive assessments and one a negative assessment of their attitudes to mathematics at school. At college the attitudes of the six heads to mathematics included two positive and two negative.

I wondered whether past attitudes to mathematics might determine the extent of active support the head would be able to give to the project. Finlay (a first school) might be particularly vulnerable since the head gave negative assessments for both periods. Another indicator of vulnerability in some schools would be that there was a higher number of teachers giving a negative assessment than a positive one. Table 2.1 shows four different types of vulnerability.

1. There were four first schools, Fleet, Foster, Fowler and Finlay, and three middle schools, Melia, Missingham and Makewell, where there were more teachers with negative than positive attitudes during their own schooling.

2. At two schools in each phase, Foster, Finlay, Melia and Makewell, the teachers gave an average assessment of D for their attitudes to mathematics at school (calculated on the basis: $A = 2, B = 1, C = 0, D = -1, E = -2$).

3. Foster and Finlay, Melia, Missingham and Makewell had a higher number of teachers who gave a negative assessment than a positive one of their reactions to mathematics at college.

4. The teachers at Melia, Missingham and Makewell had an average assessment of D for their attitudes while at college.

The schools which had a higher number of teachers with negative than with positive attitudes to mathematics at both school and college might require special help at the support visits. The schools in this category were: Foster, Fowler, Melia, Missingham and Makewell.

Constructing the questionnaire

The questionnaire was based entirely on interviews I conducted during the spring term of 1976. At each school the head, the co-ordinator and two teachers were interviewed individually. For these interviews the head had been asked to nominate the key teacher with a more negative attitude to mathematics and another teacher with the most negative attitude to this subject on the staff. Not one of the heads appeared to have any difficulty in nominating these. I realized that the selection would be subjective, but I thought it

Table 2.1: The total assessments made by the teachers at each school of their attitudes to mathematics at school and college, and to teaching the subject.

Scale: A positive – E negative

First schools

		School					College						Teaching				
		A	B	C	D	E	A	B	C	D	E	Untrained	A	B	C	D	E
area 1	Frame	3	4	4	(3)	0	3	4	(4)	1	0	2	2	(11)	1	0	0
	Fleet	0	3	(5)	4	3	0	6	6	(3)	0		2	(9)	3	1	0
	Foster	1	0	(3)	3	3	1	2	(2)	4	0		1	(4)	3	2	0
area 2	Flanders	0	(3)	3	2	1	0	(4)	5	0	0		0	(6)	2	1	0
	Fowler	1	4	(7)	3	3	1	(2)	10	2	2	1	(1)	11	3	3	0
	Finlay	0	1	4	1	(5)	2	4	2	(3)	0		(7)	3	1	0	0
	Totals	5	15	26	16	15	7	22	29	13	5		13	44	13	7	0
		20			31		29			18			57			7	

Head included in assessment where numerals are encircled

Middle schools

		School					College						Teaching				
		A	B	C	D	E	A	B	C	D	E	Untrained	A	B	C	D	E
area 1	Melia	1	1	3	⑤	5	0	1	④	10	0		1	④	8	2	0
	Meakins	③	3	5	2	1	2	③	6	2	0		3	⑨	2	0	0
	Missingham	0	4	③	3	3	0	1	4	⑤	1	1	3	⑤	4	0	1
area 2	Movehall	0	4	④	2	1	1	⑤	4	1	0	2	3	⑥	2	0	0
	Makewell	②	0	4	4	3	0	2	3	⑤	3		①	4	4	4	0
	Measures	7	③	5	2	1	2	7	⑥	3	0		⑤	10	3	0	0
Totals		13	15	24	18	14	5	19	27	26	7		16	38	23	6	1
		28			32		24			33			54			7	

Head included in assessment where numerals are encircled

important to discover the attitude to mathematics of those teachers the head regarded as having a negative attitude. The interviews were based on the following schedule.

1. 'Many teachers dislike mathematics. How do you feel about this subject?'

2. If the teacher expresses dislike: 'When did you first begin to dislike mathematics? At school? Tell me about this.' (I also tried to find out whether the cause was a teacher, a topic, absence from school, home expectations, the pace, a textbook – and which aspect caused most trouble.) If the teacher expressed a liking for the subject, I would explore this further.

3. 'How did you get on at college? Was the course in mathematics useful? How long was it? What did it include? How could it have been improved?'

4. 'What do you feel about teaching mathematics?' ('Confident?' 'Insecure?' I did not use these words during the interview unless they were mentioned by the teacher.) 'So what do you do in the circumstances?'

5. 'Have you done any mathematics since leaving college?' (By reading or attending courses?)

6. 'To whom would you go for help in mathematics if you needed this?' (A friend? A colleague? The co-ordinator? A book. Which one?)

7. 'Are there any changes you would like to make in the teaching of mathematics with your class? How can we help?'

8. 'At the five working sessions we plan, would you kindly make a note of times when you feel more comfortable about mathematics and of times when you feel frustrated.'

All the statements made during the interviews were recorded in writing because the teachers were not willing to have the sessions tape-recorded but did not object to notes being taken. Unless a statement repeated exactly that of another teacher, it was included. I had no intention of establishing a measuring instrument, only of constructing a table which would roughly categorize attitudes on a five-point scale.

Responses at interviews

It was interesting that no teacher interviewed, even those with the longest experience, appeared to have any difficulty about recalling his or her attitude to mathematics at school. However, some experienced teachers said that they remembered nothing at all about any professional course in mathematics. It was possible that even the assessments made by the teachers of their attitudes to mathematics while they were at school were of doubtful reliability, but this was what they said.

When I studied the statements made by the heads and teachers interviewed, it seemed that those made about attitudes to mathematics while at school were more negative than positive. I had hoped that by including heads and co-ordinators I could avoid an overall negative bias in the statements. What I had not expected was that all but one of the first school heads and half of the middle school heads would have had negative or neutral attitudes to mathematics while they were at school. Moreoever, some of the co-ordinators also had negative or neutral attitudes to this subject at school. In view of these circumstances the negative slant of many of the statements made by the interviewees about mathematics was not surprising.

After the interviews, all the statements made about attitudes at the three different stages were arranged in random order in the appropriate sections to form the draft questionnaire. In order to select the 15 statements in each section which correlated best with all the statements taken together, the draft questionnaire was administered to teachers in non-project schools (selected by the mathematics adviser) from areas resembling the two areas in which the project schools were situated.

The scores for each item were totalled from the completed questionnaires. Reliabilities and correlations were analysed by a computer program. In general the 15 statements selected for each section were those with the highest correlation coefficients but the need for variety was also taken into consideration.

Questionnaire I Attitude to mathematics while at school.

Please tick one column only for each statement.

1. Strongly agree   2. Agree   3. Neutral   4. Disagree   5. Strongly disagree

	1	2	3	4	5
1. I got more and more behind because I had a teacher who was not concerned about those who found the subject difficult.					
2. I always liked the subject because I had teachers who encouraged me.					
3. I remained wary of mathematics until I left school. I convinced myself that I was a failure.					
4. I was bored stiff with mathematics at school and was not exhilarated at all.					
5. I went through school taking little part in mathematics lessons because the teacher had little interest in the ones who did not understand.					
6. At secondary school I soon began to feel a failure at mathematics.					
7. I loved mathematics at school.					
8. Mathematics became confusing as wider aspects were covered.					
9. I did a traditional course and liked it.					
10. I was put off by mediocre teaching in mathematics.					
11. I loathed mathematics at school and had a block which came through being petrified.					
12. I disliked mathematics at school. Algebra and geometry were incomprehensible.					
13. I was bewildered and puzzled by mathematics when I was at school.					
14. I had many changes of teacher and did not understand mathematics.					
15. I have always loved mathematics and was good at it.					

Questionnaire II Professional training in mathematics at college.

Please tick one column only for each statement.

1. Strongly agree   2. Agree   3. Neutral   4. Disagree   5. Strongly disagree

	1	2	3	4	5
1. I liked the course – it gave me a chance to think.					
2. Considering the time allocated, the course at college was reasonable and gave an idea of what things might be like in the classroom.					
3. I enjoyed the course but learnt nothing positive to help in the classroom.					
4. The professional course was poor and did not help in the classroom.					
5. The professional course was of little interest or value. We were given some activities but no attempt was made to show their purpose.					
6. The course was not particularly good as judged by classroom needs.					
7. My college course was excellent. It began at the beginning and proceeded in a practical way.					
8. The professional course was good.					
9. The lecturer was enthusiastic but did not do anything to improve my confidence.					
10. The course was of little use and I skipped it whenever possible.					
11. The course was interesting but not of much use.					
12. Mathematics was a delight at college because it was geared to classroom method. We handled material ourselves.					
13. At college we were shown all the equipment but did not have time to use it.					
14. The college course was a washout.					
15. The course at college was airy-fairy.					

Questionnaire III The teaching of mathematics.

Please tick one column only for each statement.

1. Strongly agree   2. Agree   3. Neutral   4. Disagree   5. Strongly disagree

	1	2	3	4	5
1. I am confident in teaching mathematics.					
2. Mathematics is a difficult subject to teach.					
3. I am not confident about all the mathematics I have to teach so stick to basics.					
4. I like teaching mathematics which I think is an exciting subject.					
5. I have become confident in teaching mathematics and very few of my children dislike it.					
6. I am not very happy teaching mathematics. I should like to know that what I am doing is right.					
7. I realize the importance of mathematics but do not enjoy teaching it.					
8. I am interested in mathematics but I am not confident in teaching the subject.					
9. I am not confident in teaching mathematics to mixed ability groups.					
10. I am not short of ideas so do not depend on books when teaching mathematics.					
11. I have developed a guilt complex about teaching mathematics and now understand why I had difficulty at school.					
12. I am not confident in teaching mathematics so I rely on a textbook.					
13. I do not think I teach mathematics as well as I could because I do not know what to teach nor why.					
14. I like teaching mathematics and feel confident when teaching the subject.					
15. I like teaching mathematics because I am interested.					

Comments made during the interviews

In each of the three sections of the questionnaire, the wide range of the statements made by the teachers at the interviews was illustrated. For example, in the school section (Questionnaire I) there are some statements which express the teachers' enjoyment of mathematics at school, but more which express their feelings of failure, boredom or confusion. In the college section (Questionnaire II) one teacher thought his course was excellent because of its practical nature, while others rated their professional courses as of little value, even when the course was interesting. The comments in the teaching section (Questionnaire III) show a similar range, from confidence to the development of a guilt complex. Not many of the statements are neutral. However, the questionnaire cannot provide evidence of attitude change on the part of any one teacher. I am therefore citing statements made by four individual teachers, two from each phase, in the course of the original interviews. The teachers' self-assessments at the three stages are included at the end of what each says.

A deputy head of a first school said:

I left school at the age of 15 and went to a technical school where I did no mathematics. I had hated mathematics at school . . . At college I was interested but not confident about teaching mathematics. Although the course was at child level and the students were interested, there was very little for infant teachers. I am interested but not confident about teaching the subject. (E, C, C)

The statements made by the next first school teacher illustrate the major changes in attitude which occurred during her education. Her assessments should not come as a surprise. She said:

Infant school: happy play, but little link with reality except for learning to count things and tell the time.

Junior school: Sudden pressure to produce results – table tests, learning how to do sums, etc. I withered under the competitive spirit which built up, as I was a relatively slow learner. I enjoyed learning by rote and off by heart; it was the only way I could keep my end up. Enjoyed everything familiar, feared any new ideas.

Secondary school: Had bad or fearsome teachers up to the age of 15, hated maths. But an older, more understanding teacher in the O level year suddenly brought it to life, by explaining the uses of things we'd learned and by giving us individual help. She gave us things to find out and games to play using maths and this gave us another aim to work for apart from exams.

College: Absolutely hopeless! Just played with (coloured) rods.

Teaching: Just do what I know from my own schooling and have found useful. (C–B, E, C–D)

Perhaps the most interesting feature of this account is the change in attitude from the early enjoyment of learning by heart to the rehabilitation caused by seeing a purpose in things learned and taking an active part in new learning.

A middle school co-ordinator who had had an unsatisfactory mathematical education improved her background knowledge by attending two local courses; she said:

I enjoyed maths at primary school but did not understand at secondary school . . . I had many changes of teacher at school and only one of these was good. At college I had a term of 'methods' from an education lecturer. I have been to a number of courses on middle school work . . . I should do far more practical work but this term I have not got organized. (D, C–D, A)

A middle school co-ordinator who proved to be an outstanding teacher of children of all ages commented:

I lost the magic of mathematics at the secondary stage. At the beginning I liked it; then it became so exam-structured that I disliked it. This was the teacher's fault. At college . . . we handled material ourselves. I enjoyed it; it was a delight. As a young teacher I was helped by a very good head. I also compared notes with my sister who began to teach infants at the same time. I no longer believe in showing children how to do things on the board . . . I rarely use a textbook because this takes me away from the children. (C, B, A)

I was grateful to the heads and all the teachers interviewed for their frankness. The interviews had another advantage. They gave me the opportunity to meet and talk, in a less formal situation, with many of the teachers who would be attending future working sessions. Although I had planned the form of the interviews, the atmosphere was relaxed and informal. I was able to gain an impression of each teacher's mathematical education.

The revised questionnaire

It was unfortunate that the revised questionnaires were not ready to be sent to the project schools until the first input of in-service education was completed. This was due partly to the lengthy first draft and partly to inadequate duplicating services at the LEA office. More than two terms elapsed between the initial assessments made by the teachers and their completion of the revised questionnaire. The teachers' total scores for each section of the questionnaire were calculated from their estimates of their attitudes to each statement. These scores were compared with their original assessments. The percentage of teachers for whom the questionnaire scores and their original assessments were in agreement was only about 40 per cent. The discrepancies could have been caused by any one of the following factors:

1. the time lapse between the assessments and the completion of the questionnaires;
2. the effect of the first input of working sessions and support visits;
3. the difference in the nature of the assessments and the questionnaire.

Many teachers said that they found it more difficult to make assessments of their attitudes on a five-point scale than to express the extent of their agreement with the statements in the questionnaire. But it is difficult to understand how any circumstances could alter the teachers' assessments of their attitudes to mathematics at school or at college. I therefore decided not to proceed further with the comparisons. Nevertheless, the exercise was valuable in that I had had personal contact with two-thirds of the teachers I would be working with, and I knew some of the problems they had encountered in the past with mathematics. Moreover, I had been alerted to those schools in which there were more teachers with negative attitudes to this subject than positive. This knowledge stood me in good stead during the first input.

But there were other factors which could have affected the teachers' attitudes, and these should be kept in mind. As pupils themselves they had had no systematic introduction to practical activities and the appropriate language patterns before symbols were used. In consequence, many teachers relied on their memory of processes (such as long division and multiplication of fractions) rather than on an understanding of the concepts involved. Few of the teachers in project schools had liked mathematics when they were at school; many had disliked the subject. Moreover, it was clear from the questionnaires that for many teachers the professional course in mathematics did little to change their adverse attitude to this subject.

In Chapter 3 other outcomes of the preliminary visits to the project schools will be recorded: the introductory visits, the interviews with selected children, and the observation visits.

Preparation for the first input

Introductory visits to the heads and the teachers

At the first visit the mathematics adviser introduced me to the head. I outlined the aims and proposed design of the project first to the head and then to the teachers. After discussion with their staffs, all the heads expressed their willingness to participate in the project.

At subsequent visits I ascertained from the heads:

1. the type of organization in the school: mixed ability groups, vertical groups or streaming (and whether this organization also applied to mathematics);
2. the degree of freedom allowed by the timetable;
3. the scheme for mathematics; how it was prepared and who was responsible;
4. the textbooks or workcards in use.

I also met the co-ordinators. I asked the heads and the co-ordinators in the off-site schools to nominate a team of two or three key teachers who would be released, with the co-ordinators, to attend the working sessions at the teachers' centre. Ideally, key teachers should be drawn from different year-groups in the schools, in order that they could influence their immediate colleagues as far as possible. I emphasized that key teachers should be those to whom their colleagues would listen.

Interviews with selected children

The head and the teachers were asked to nominate three children from each year-group: one able, one from the middle of the class and one slow-learner. There were 108 children altogether, 54 from each phase, and I interviewed each child individually with two main objectives.

1. To check the findings of my previous experience:
    a. that teachers are more anxious about number than about any other aspect of mathematics;
    b. that teachers avoid certain practical activities altogether, for instance when dealing with fractions, capacity and volume;
    c. that many children have an inadequate knowledge of number facts for the written calculations they are expected to do.

2. To discover the attitudes of some of the children to mathematics. This knowledge would help me in planning the content of the working sessions.

The interviews, therefore, took the following form.

1. I asked each child about his family and his likes and dislikes within the school curriculum.

2. I gave each child practical activities:
    a. in fractions: finding one half, one quarter of a length of ribbon, a glass of water (other identical glasses were available), and a lump of clay (balance scales were to hand). With the younger children the language was simplified;
    b. in comparing the sizes of two rocks.

3. I questioned each child orally to find the extent of his number knowledge. Where appropriate I went on to written calculations. I found that:
    a. for none of the children in first schools was mathematics their favourite subject, although two boys said they 'liked doing sums best of all the things we do at school'. Six children at middle schools said that mathematics was their favourite subject. However, two able boys said that mathematics was definitely not a subject they liked 'because it is so dull';
    b. none of the first school children was intimidated by the practical problems. On the contrary, many began to use the materials as soon as they sat down. But their use of materials showed how unfamiliar they were with the activities. For example, when comparing the sizes of two rocks all the

children began by using balance scales. Twenty-one of the 54 first school children pointed to the upper pan as the heavier (eight of these were 7 or 8 years of age). The teachers of these children volunteered that they had not provided weighing experiences except as 'play'. The teachers commented in a similar way about 'water activities'; these, too, were regarded as play.

At the middle school stage, only three children confused heavier and lighter. Once more, many teachers admitted that they did not provide either weighing or capacity activities for the children.

The activities giving experience with fractions produced some interesting results. Thirty of the 54 children from first schools completed all three tasks of halving quantities, although some of the middle school children found these tasks difficult. Halving a length of ribbon caused most problems; many children guessed the position of the middle and took time to decide how they could check and adjust their estimate. All said that they had never been given activities like this before. Their comments were confirmed by their teachers, who said that they relied on textbooks for the work on fractions and usually demonstrated the only practical activity suggested (dividing a circle of paper, sometimes referred to as a cake, into equal parts).

c. the children's oral knowledge of number facts spanned a wide range. Only one third of the older children at first schools had speedy recall of the facts they were using in written calculations. Well over a half of the third and fourth year children had a most inadequate number knowledge, although some showed great ingenuity in using the few facts they knew to obtain other facts. No first school child could carry out correctly $62 - 26$ as a written calculation, although the teachers included examples of this kind in their teaching.

At the middle schools, 20 of the 54 children had a good knowledge of number facts; 13 had far too slender a knowledge to be able to tackle the written examples they were being set. One third were able to carry out a subtraction of the type already mentioned and to explain, with understanding, what they were doing.

These interviews yielded some valuable evidence to help me decide on the practical work which should be included in the first and second inputs. It seemed clear from the children's performance, and their description of mathematics in school, that lessons consisted mainly of working by themselves from textbooks or workcards. They did not mention activities. Certainly there had been little practical work in fractions. I decided that this topic must be included in the first input. Providing practical problems in fractions would also give me an opportunity to revise the situations (and the language patterns) which give rise to one or other of the four operations. The children were evidently unfamiliar with these situations. Moreover, their vocabulary did not match their mathematical thinking; for example, they referred to big and little when they meant tall and short.

It would also be important to help the head and the teachers to prepare and try out a mathematics scheme. Many teachers seemed to have little idea of where the mathematics they were doing was leading, or of what the children had done previously. The contacts I had had with the teachers concerned, after the interviews with the children, had also been valuable and had given me further insight into their views of what they were doing in their classrooms.

Observation visits to teams of key teachers

The purpose of the preliminary observation visits was to establish a baseline for the range of teaching styles used during mathematics sessions in each of the project schools before the first input. The three major teaching styles I expected to find were class teaching, group teaching and individual work (and many different combinations of all three styles). At one extreme, class teaching could mean class instruction; at the other, it could be a lively exchange between the teacher and the pupils during which the pupils made a major contribution. In the same way, individual work could mean that each child was working from the same scheme at his own pace and had little opportunity for discussion either with the teacher or with his peers. (Systems of commercial workcards for mathematics were often used in this way.) On the other hand, in addition to allocating individual work, the teacher might hold regular group sessions to focus attention on common problems as well as on original solutions.

I was aware that an objective and detailed assessment of the teaching of mathematics in individual schools, even limited to the teaching methods of the co-ordinator and key teachers, could be assembled only as the result of several visits to each school. There were certain features in the teaching of mathematics which could indicate the methods used by the teachers and could be perceived in advance of a formal observation visit. These included the seating arrangement of the children: were the tables or desks arranged in groups or rows? Did the teacher more often talk to the whole class from one point in the classroom, such as the blackboard, or move from group to group or child to child? Was there any kind of mathematical material on display? Was this prepared by the teacher, or was it children's work? Was the display attractively presented? Was the equipment sufficient, in good condition and readily available? Were there any mathematics books of general interest available? I noted as much as possible when I was first taken round the schools to be introduced to the co-ordinators and key teachers, and on subsequent visits when I interviewed selected children and teachers. By the time of the observation visits I had already spent four or five days at each school (three for interviews). I was particularly anxious to gain an impression of the range of teaching styles used by the key team before the observation visits. I anticipated that nervous teachers might vary their normal practice when being observed. There were other ways in which I hoped to gather the information I needed. During the interviews I asked each teacher for her view of the teaching methods she used.

I was able to check my observations with the heads. At nearly all my subsequent visits the heads volunteered information concerning the co-ordinators and key teachers, their confidence or lack of it and where they needed help.

The first observations

In the event, the observation visits were carried out one year after the reorganization of the schools in September 1974. All the schools except one had suffered from a high staff turnover (about 50 per cent in one year) as a result of the reorganization. In addition, seven of the twelve heads were newly appointed (three at first schools and four at middle schools) and in their first headships. The teaching of three of these had been in other phases than those for which they were now responsible. Moreover, all the schools had some teachers (experienced in schools for older pupils) whose inclination was for formal methods. Furthermore, every school had some teachers who were in their first posts. One first school and two middle schools were without mathematics co-ordinators during the early stages of the project.

One school only, a first school with a new head, had an up-to-date mathematics scheme (prepared by the head herself). The mathematics adviser had been dismayed to find that so many schools in the borough had neither mathematics schemes, nor a copy of the guidelines prepared a year or so before 1974 by a group of teachers in the borough. On her recommendation a number of first and middle schools in the borough had adopted one of two new commercial mathematics schemes. Unfortunately, the co-ordinators had not at that time been trained for their new responsibilities, and were unable to give informed advice in the use of the new schemes. Consequently, these were used in a restricted way. Not unnaturally, teachers to whom the schemes were totally new followed them closely and not to the best advantage of the children. At one first school second year children were using first year workbooks, because the teachers felt that the content was so novel.

During the preliminary interviews and observation visits I acquired useful information about the mathematical background of the heads and some of the teachers. The heads of Fleet and Foster had a good knowledge of mathematics. Those of Foster, Meakins, Makewell and Measures, and the co-ordinators of Melia, Movehall and Measures had attended mathematics workshops I had run previously. I therefore expected that my aims for the project, although rather different from those for earlier workshops, would not come as a shock to these heads and co-ordinators.

There was one unexpected feature. I had asked the heads to nominate key teachers in different parts of the school whom their colleagues would respect and to whom they would listen. In the event, half of the key teachers were in their first posts. However, when I questioned the heads it became evident that all had given careful thought to the nominations. But I soon found out that some of these young teachers were struggling to control their classes, and that until they had overcome these problems, they would not be ready to pass on to colleagues in their own year-groups what they were learning at the working sessions. Moreover, some experienced teachers would not be willing to listen to such young teachers.

Most of the schools lacked basic equipment for teaching mathematics. All the co-ordinators volunteered that their first task would be to review and augment existing equipment. There were few attractive displays of mathematics material and there was hardly any children's work on display. Although some schools had been working on projects which integrated several aspects of the curriculum, mathematics was never included.

In the first schools, all the classes were unstreamed, and there were no sets for mathematics. Of the nineteen key teachers, all except three appeared to teach mathematics formally, even though, in other aspects of the curriculum, the work was often less formal. There was little discussion during mathematics sessions, either between the teacher and the children or among the children themselves.

All the classes in the middle schools were similarly unstreamed, but the third and fourth year children were usually allocated to sets according to their ability in mathematics. This form of organization had the advantage of reducing the number of children in each set, particularly the slowest sets.

The head of Movehall had introduced vertical grouping over a two-year age span, together with team-teaching. However, for mathematics all the classes were regrouped according to mathematical ability. The number in each group was low, but the time allocation was about half the usual allowance of four to five hours a week.

Four of the twenty key teachers in middle schools were successfully teaching in an informal manner. Twice that number said that they were formal teachers by inclination. The other eight were attempting to adopt a less formal style. There was little discussion, except in groups where activities were provided.

Note: The tentative observations in which teaching methods are sometimes referred to as formal or informal are merely descriptive and are not meant to imply any criticism. It was important for me to note those teachers who were not providing activities or opportunities for discussion, so that I could arrange for them to have these experiences if they had lacked them in their professional training. In this way my observations would help me to plan the first support visits.

Chapter 4 describes the first input of working sessions and the teachers' reactions.

The first input: the centre-based working sessions and the teachers' reactions

Organization and content

The working sessions and support visits which made up the first input took place during the summer and autumn terms of 1976, one term later than was originally planned. The summer was not a good time to begin a project: there would be disruptions caused by the numerous school journeys and sports days which take place during this period. Moreover, the long summer break which followed would give the teachers a chance to lose a certain amount of enthusiasm and to forget what had been covered. However, in order not to lose the impetus generated by my preliminary visits, the advisers and I decided to organize five working sessions during the summer term and to begin the support visits.

The first input was part of the whole plan to bring about changes in the teaching of mathematics. This would involve all the teachers who attended the working sessions in careful planning and the subsequent trial of sequences of activities. It would also mean a change in teaching style for those who had relied solely on class teaching techniques for mathematics. The working sessions were primarily intended to provide sample sequences of activities to help children to acquire certain concepts; to emphasize the language patterns associated with the various situations which give rise to the four operations; and to increase the teachers' knowledge of mathematics.

I realized that the changes I hoped the teachers would try to make would be facilitated by giving them experience of learning at their own level by means of investigations and questioning. Moreover, this way of working should increase their own understanding of important concepts and build up their confidence. If they worked in a group with their colleagues they would experience, at first hand, the exhilaration of solving problems and of comparing and appraising the different methods they used.

Two types of organization were planned for the working sessions. One first school and one middle school in each area had their sessions at the school, and the head and all the teachers attended. (In addition, in one area, the key team from the high school attended the working sessions at the middle school.) The remaining two first and two middle schools in each area sent key teams of teachers to the teachers' centre for the working sessions. The key team from the other high school attended the sessions for that area at the teachers' centre.

I hoped to discover which was the more effective method of conducting working sessions: that involving the head and all the teachers at individual schools or that for key teams at the teachers' centre. The latter had the potential of influencing four or five schools if, as I envisaged, the key teams were able to pass on the procedure and content of the working sessions to their colleagues. The relative effectiveness of the two patterns would be assessed in terms of the actual changes made in their classrooms by all the teachers in each school.

The working sessions were originally timed to take place once a week in the afternoon and to last for three and a half hours. But some of the teachers were too tired to concentrate after school hours and the duration had to be curtailed. Moreover, the sessions took place during the hot summer of 1976 in hutted accommodation, which added to the discomfort.

The overall theme of the working sessions was pattern. The programme and the papers for circulation were planned in consultation with the advisers.

1. The teachers would be organized in groups (of their own choosing); towards the end of the first input the value of working in this way would be made explicit.

2. The prevailing tone of the sessions would be encouragement, to help the teachers to see that children, too, should be encouraged.

3. There should be as little 'lecturing' as possible by any of the staff.

4. The programme would comprise:
   a. planned activities for various concepts, using the appropriate language patterns;

b. games at adult level which could be adapted for children of different ages;

c. planning and trial by the teachers of a series of activities on chosen topics;

d. discussion of types of organization which would facilitate working in groups;

e. discussion of any activities tried with children between the working sessions;

f. appraisal of the working sessions.

5. The papers distributed at the end of each session would serve:
   a. to remind the teachers of what had been covered;
   b. to relieve them of the need to take notes;
   c. to help absentees to find out what had been covered;
   d. to help the teachers to understand the intended structure;
   e. to provide a specimen sequence of activities to assist teachers with their own planning in the future.

6. The content, all of which was developed by means of activities and accompanied by the appropriate language patterns, would comprise
   a. the minimum number knowledge which should be required before children are given extensive practice in written calculations;
   b. a variety of situations which give rise to the four operations and the appropriate language patterns;
   c. different methods of calculation to emphasize the importance of helping children to develop more than one method;
   d. place value and its extension to fractions and decimals (which would provide opportunities for revision of the language patterns of the four operations);
   e. a summary of a variety of number patterns (for teachers at upper middle schools and lower high schools).

Each of the papers to be distributed would include 'practice' examples to help teachers to become familiar themselves with alternative methods of carrying out calculations, and to show them that paper and pencil investigations, designed to give children experience in finding number patterns, could also provide them with quick practice in mental calculations.

Progress of the working sessions at the teachers' centre

I was surprised to find that the mathematics co-ordinators were so young: all were under 30 years of age, and some were nearer 25. Several of the key teachers were even younger; some were in their first or second year of teaching. Since they were expected to influence their colleagues, I wondered whether too little weight had been given by the heads to the status of the key teachers in their schools.

Most of the teachers knew only colleagues from their own schools even when corresponding first and middle schools were on adjacent sites. Many pairs of teachers from adjacent schools settled down to work together. (This had been my main purpose in organizing working sessions for the age range 5 to 12 years.) The three high school teachers were known to few, if any, teachers from middle schools.

The teachers worked well on the activities introduced, but soon revealed how unfamiliar they were with the variety of situations which required the use of one of the four operations for their solution, or with the appropriate language patterns. The programme, which included frequent revision of these concepts in different settings, was clearly justified. I had intended to lead a discussion, on the papers distributed, before the end of each session, but the teachers were too tired by then to concentrate.

As I expected, there was considerable resistance to teaching children more than one method of written calculations, particularly for subtraction. The teachers had to agree that adults used many different methods for mental calculations, since we had collected six methods for the mental subtraction I gave them, but they thought that children would be confused if they learned more than one method. I realized that they would not change their minds until they were thoroughly familiar with these methods themselves. We had frequent practice sessions which I hoped would show that I, too, accepted the need for practice.

At each of the three sessions after the first, a variety of number games were played at adult level. The need for the teacher to work with the group of children to whom she was introducing the game, in order to ensure that the children understood the purpose and developed appropriate strategies, was emphasized from the outset. The further necessity of following up a sequence of games with quick oral work to assess learning, and of reinforcing the children's understanding of the purpose of the games, was also emphasized. Teachers were attracted by the games, and developed their own versions for the children they taught. These were often the first activities which the teachers tried with their classes between the working sessions. Despite the caveat I had given, all the teachers expressed surprise at the extent of supervision required when the children played a game for the first time. It was also evident that some teachers regarded games as fringe activities rather than as valuable tools for memorizing essential number facts. However, before the support visits began, the games were the first step towards less formal methods of teaching.

Assessments made by some teachers

During the fourth session with the teachers from the area of social priority (area 2), a situation arose which changed the subsequent programme for all the teachers. The co-ordinator from one first school sat at a table by herself. At the tea-break I asked her whether the working sessions had helped her in any way. Her reply was:

> No. You've totally confused me. Your methods are all wrong. You teach us a game and then ask us to adapt it for our children. How do we know?

Realizing how strongly this teacher felt, I explained to the whole group that some teachers had not found the working sessions helpful so far. I therefore asked them for written comments about the working sessions, and suggestions for making these more helpful. Some of the replies from the group now follow, beginning with that of the first school co-ordinator, who had followed up her initial protest by saying that she had always hated mathematics and that she would not have accepted the post of co-ordinator except that no one else would take this on. She wrote:

> I have found the course confusing. It has left me with little knowledge of where what fits in as the course tends to jump about quite a lot from various age groups.
> I would like to see a basic course covering all maths concepts, e.g. number, time, weight, volume, measure, fractions, etc., slotted

into a logical pattern of progression. Questions like where does one introduce time in relation to other maths concepts and where does one leave it and take on something else and then when one should return to it.

One of the two key teachers at the same school wrote:

> The aspects of maths covered in the course seemed geared more to middle and high school level. Therefore I have found little of value to help me as a first school teacher. Many of the ideas are those which I think most of us have already covered in our college training as teachers, though it is important I think to have been reminded of them. I would have preferred a topic, e.g. measuring, to have been begun at first school level and carried through to middle and high school stages, instead of beginning at what seems to me an intermediary stage and then asking what could be done at first school level and higher stages.

(This teacher was in her second year of teaching.)

Key teachers from the other first schools expressed similar doubts and requests. One co-ordinator who assessed her own teaching as formal wrote:

> It makes me feel that we should have guidelines in maths ideas. I feel in some areas I have missed things and with more guidelines everybody would feel more confident . . . I have picked up some useful ideas.

(Guidelines for the borough had been distributed three years earlier.)

A key teacher in her third year wrote:

> I think I would have preferred a course for first schools only and middle schools only and then a couple of meetings together as a lot of time has been taken up with maths for older children and therefore irrelevant. Made me more aware of what is lacking so far in my maths teaching in practical ways.

A key teacher in her second year said:

> I like many of the ideas, especially games for the classroom.

Because the emphasis so far had been on the development of an understanding of counting, of the operations, and of place value, I had not been aware that first school teachers had felt that many of the activities were more appropriate for middle schools. Perhaps this was a valid charge, since the comments of the middle school teachers were more appreciative, as the records illustrate. One teacher wrote:

> I have found the course very helpful. This is my first year of teaching and I have found that I can cope adequately in the classroom, in that I can teach number quite easily, but I needed to know more practical activities suitable for the age group I teach (first and second years). It has helped me see many different ways of tackling particular things.

A probationary teacher of third years who was having problems in maintaining order was less confident:

> The course has given me a considerable number of ideas which will be of great help when introducing children to the first stage (of practical activity). However, I feel rather confused as to when and how to get the children to write up the work and to do written practice.

An experienced teacher from the same school (also having trouble in maintaining order) wrote:

> The course has been interesting and instructive, and I have found, personally, that I have had to *think* hard on my approach to teaching maths – also I realize the difficulties encountered by the children.

From the same middle school an experienced teacher wrote:

> I am *more* aware of the need for more practical work, and the need for careful use of vocabulary, but it's only really confirmed what I knew.

A fourth teacher at this school, a graduate in his second year of teaching, wrote:

Helpful in that it has made me aware of the vocabulary I use in teaching mathematics and consequently I realize I need to be more consistent in the language and terms I use. It has also enabled me to see processes that I take for granted in a fresh way and to become much more aware of the difficulties some children face in basic number work. It has been useful too in seeing progression in learning maths. Rather daunted by the amount of practical work needed . . .

In the fifth session for the other group I began by asking the teachers for a written assessment of the course so far, explaining that I was requesting this earlier than I had anticipated because the teachers at one school had been critical of the sessions. If these views were general, I would like to reconsider the structure and content of the sessions on their advice.

Teachers at first schools

An experienced teacher who subsequently changed the balance of her lessons considerably wrote:

I have enjoyed the course, especially discussions about games and number . . . Interesting too hearing comments of other teachers. Only drawback – I really find (myself) a bit at sea at times . . .

A key teacher, informal in her methods, who was a deputy head wrote:

All activities have aided my understanding.

Another key teacher, in her first post, who had made a good beginning at the same school commented:

I have learned a great deal. I'm particularly pleased that you have concentrated on number as there are so many ideas available for other mathematical concepts like weight, volume, length, etc. It would be helpful to have ideas of activities that children can do without supervision that are useful (i.e. not just 'sums'). What the course has given me is an approach — to get away from rigid adherence to [the workbook series in use]. I only hope it lasts.

(This teacher left before the second input to become a mathematics co-ordinator at another school.)

Teachers at middle schools

The co-ordinator of one school began by listing the aspects she had found most useful, and then wrote:

> Generally, I feel that the course has been useful from the point of view of clarifying my aims with some topics. Also I have been more aware of the methods children themselves use to solve problems.

A key teacher in her first post at the same school (teaching slow nine-year-olds) commented:

> Whole course most useful, ideas and practical activities; although I am perhaps biased since most of the activities have often been related to the age and ability of my group. A great help to make me notice and realize the way in which children work things out for themselves. Method. Amount of calculation that goes through their minds. Unusual things they spot/pick out in a situation.

I was reassured to know that the working sessions had influenced these teachers to observe and listen to the children, and to become appreciative of the children's ability to think things out independently. It also seemed that these teachers were successfully adapting the activities used at the working sessions for the children they taught. I was left wondering why the first school teachers found this kind of adaptation so difficult.

At another school, a key teacher in her first post who had already tried many ideas from the working sessions with her own children wrote:

> I have found most of what we have done most interesting. It has been useful to use some of the games in the classroom . . . The course has helped me to be more critical of my own methods of teaching maths.

This young teacher was most successful in transmitting ideas from the working sessions to other teachers in her year-group, and in encouraging them to adapt these ideas themselves.

The comments made by two teachers of pupils in the first year of the high school, both in their first posts, were indicative of their views of mathematical investigations:

> Course has proved interesting and useful to *me* but the work involved can rarely be used in upper school due to pressure of exams and time. I would definitely make use of it teaching remedial groups at the lower end of the school (which I am not doing at present).

The other teacher taught a remedial class. He wrote:

> I regard informal teaching as demanding *far* more careful structuring by the teacher; with *more* time on preparation (and correspondingly less time on marking).

After receiving their comments I described to this group the corresponding session with the parallel group and their general request for more help in structuring sequences of activities. The second group then decided that they would prepare a sequence of activities on subtraction, covering the age range 5 to 13 years.

They began by working together, everyone contributing by suggesting activities and games for different stages. They then broke into smaller groups for the final ordering and recording of the material. This sequence was duplicated and distributed to key teachers in both groups and to teachers in the four schools receiving the other pattern of in-service education. Heads and co-ordinators received the entire sequence of activities; key teachers received a limited section only, so that they should not be daunted by the extent of the material. (The final document was distributed to other schools in the borough on request.) For the first time the key teachers began to experiment with the introduction of activities other than games, with the help of the subtraction scheme.

The first group warmly welcomed the scheme and spent the fifth session in discussing the activities included and in preparing the materials they would require when introducing these to their classes.

Many of the teachers expressed a determination to experiment with a sequence of activities which would ensure progression.

Further working sessions

Because the five whole days originally planned for the working sessions had been cut to five half days, I decided, in consultation with the advisers, to offer two extra working sessions during the autumn term. For the first time, heads of schools were invited to these centre-based sessions. This was partly in response to a request from the head of Flanders, but also because I had already become aware of the advantage of having the head present at school-based working sessions. Four of the eight heads attended one or both of the sessions. In retrospect it would clearly have been beneficial to have had the heads at all the working sessions but that would have made the problem of teacher-release even more acute.

There were two other advantages of continuing the sessions after the summer vacation. First, the aims of the early sessions were reinforced and the content was reviewed. Secondly, the teachers realized, for the first time, how much they had forgotten during the eight- or nine-week gap since the previous working sessions. This provided a valuable discussion point – the amount children forget during the holidays, and ways of tackling this problem in order not to undermine their confidence.

The teachers agreed that many of the games they had learned could be used for the informal assessment of individual children. When playing games the children were unaware that they were revealing not only their knowledge of number facts but the extent of their understanding of concepts.

Although not a great deal of new ground was covered in these two sessions, an introduction to various ways of representation (including 'graphs which made themselves') served as useful revision of earlier activities. The sessions also gave teachers renewed enthusiasm for classroom experiment. I emphasized that I was not looking for an upheaval in mathematics teaching but a slow and steady change which would allow the children to take a more active part in the learning of new concepts.

Interim conclusions based on the centre-based working sessions of the first input

The attendance (over 80 per cent) was high in view of the time of year – the summer term – when most of these sessions took place. Furthermore, there was no cover for the key teachers in their classrooms.

It was a great help to me that all the teachers were so frank in their comments – spoken, as well as written. Their comments reflected my problems of maintaining a balance in the content. On the one hand, the teachers accepted their need for experiencing a wide range of activities to help them to understand mathematical concepts, particularly those related to the four operations which they had learned by rote. On the other hand, they required opportunities for planning, with other teachers, sequences of activities for the children they taught. They could not undertake the planning for any concept until they had achieved understanding of that concept (normally by investigation and discussion). Unfortunately the time available for sessions in the first input (16 hours) was little more than half that originally proposed. Time for planning sequential work for the classroom was therefore severely curtailed.

One factor which slowed down the pace of the sessions was the wide range of experience and mathematical background among the teachers. However slowly the sessions progressed the pace was too quick for the most insecure teachers, and too slow for the others. The papers distributed at the end of each session were, of course, intended to make up for this. Some teachers at the working sessions discussed the ways in which they used the papers to help them to plan work between the sessions for the children they taught.

Many teachers asked to be provided with detailed sequences of the development of all those mathematical concepts which they were required to teach. LEA guidelines in mathematics usually offer this kind of help (though not perhaps in the detail teachers think they need). But these teachers seemed unable to utilize the local guidelines in this way. This suggests that each individual teacher has to be involved in the preparation or modification of the mathematics scheme for her school if she is to use the scheme effectively.

One of the most successful sessions was that in which one area group planned a sequence of subtraction activities to cover the age range 5 to 13 years. They chose subtraction partly because teachers and children find this concept difficult and also because it had been covered thoroughly in previous sessions. Moreover, because the activities prepared were discussed and tried out by the teachers in the other area group, they, too, were successful in making use of the subtraction paper. But to have provided the teachers with sufficient opportunities to plan sequential activities – even for two or three mathematical concepts such as place value, capacity and symmetry – would have taken far longer than the time available in the first input. Such planning would have had to cover the age range 5 to 13 years and to cater for children of all abilities.

If I had decided to include more examples of such planning, it would have been more effective to organize the working sessions of the first input for first schools and middle schools separately, instead of for first and middle schools together in each area. On the other hand, separate sessions for first and middle schools would have meant postponing the opportunities for the staffs of corresponding first and middle schools to establish contact and ensure continuity in mathematical education when pupils were transferring from one school to another. In one of the two areas, continuity was achieved to a limited extent during the first input. But despite goodwill and the strenuous efforts of the head of one first school, exchange visits to classrooms were not arranged. In the other area the philosophies of the staffs of the first schools were very different from those of the staffs of the middle schools. In each case, the first school was more traditional in outlook. This difference inhibited contact, even on a social level, during the first input.

CHAPTER 5

The first input of support visits and the teachers' reactions to them

Introduction

The working sessions had been organized to give the teachers experience, at first hand, of learning mathematics at their own level by means of activities, and subsequently of adapting these, with their colleagues, for their children. The support visits were planned to give teachers the confidence to implement changes and to sustain these, by helping them with the organization of activities and discussion for their children. For many teachers this would be a new and possibly daunting experience. Since most of the activities and games in which the teachers had taken part during the working sessions were for group participation, I made it clear that I hoped teachers would be willing to try working with groups themselves.

Although all teachers have had experience of class teaching during their professional training, few appear to have had the experience of organizing a class of children in groups, of providing them with a mathematical investigation (requiring the use of material or paper and pencil) and of moving from group to group, observing, listening and then questioning to facilitate learning. Indeed, not many teachers have had the experience of working in mathematics with even one group in this way. I therefore planned to try to give all the teachers this experience during the support visits. (Only when the teachers are at home in the three main teaching styles will they be able to make an informed choice of style for a particular group on a specific topic. Teachers need to be confident in their ability to work in different styles.)

I did not expect that all the teachers would want to continue organizing groups for activities and discussion. Some teachers were already familiar with group work in other aspects of the curriculum, but many were not.

At this stage (December, 1976) the advisers had not yet arranged conferences to familiarize the mathematics co-ordinators with their responsibilities. With two exceptions, neither the heads nor the co-ordinators of first schools knew what was expected of them. Moreover, at that time, only one first school co-ordinator felt that she had an adequate background in mathematics. These circumstances tended to limit the scope of the support asked for by the heads and the key teams of the first schools; the idea was new to them and they did not know what kind of support to request. On the other hand, the co-ordinators at the middle schools seemed to have a better idea of what was expected of them; they also knew more mathematics.

At the support visits I tried to give the teachers confidence by helping them to translate the activities used at the working sessions into activities appropriate for the children they were teaching. I also provided further resources when the teachers came to the end of a sequence of activities and were unsure about the next stage. For a time, until they had gained sufficient confidence to draw on a range of resources themselves, they would require further support.

During the first input four whole days of classroom support were planned for each school. At first it was not easy to make the purpose of the support visits clear to the teachers, since no one had ever paid them visits of this kind before. Moreover, although the dates were agreed with the head and given to her in writing, she did not always remind the teachers about a forthcoming visit. At schools with the centre-based working sessions, teachers other than the key team knew even less what to expect. Although the aims of the support visits had been discussed at the centre-based working sessions, and at the working sessions at individual schools, it took time for the heads and the co-ordinators to organize my time to the best advantage of individual teachers. Those heads who were accustomed to children working in groups soon evolved an effective plan. Others often left the visits to chance, saying: 'We wanted you to choose what was best for you.' When the programmes for the visits were not planned in advance, time was usually wasted at the beginning of the day. On the other hand, some heads and co-ordinators discussed the support visits with the teachers in advance and provided a working programme which could be put into operation immediately.

Most teachers were timetabled to be with their classes throughout the day. This made it difficult for me to plan a session with each teacher and to find enough time for discussion afterwards while the lesson was still fresh in the mind. Furthermore, when lessons were of only 30 minutes duration, it was not easy to keep to the programme and there was little, if any, time for discussion and evaluation.

It was intended that support visits should be flexibly adjusted to meet the needs of each project school. Therefore, I had to assess the stage each school had reached before planning my tactics. For this reason a description of the state of mathematics teaching at each school is given first; details of some aspects of the support visits follow. At the end of this chapter the strategies used are summarized. The descriptions of the schools with on-site working sessions include brief accounts of the progress of these sessions.

First schools

Frame (centre-based pattern of working sessions)

The head described her approach to teaching as 'traditional'. She said she had inherited 'some experienced teachers who are set in their ways'. Five of the thirteen teachers had been trained overseas in a formal tradition. Two teachers who had been trained to teach at the secondary phase had recently returned to teaching. One of these had been nominated to be a key teacher.

Until the start of the project, the head had not given much thought to mathematics. She did not believe that understanding was essential to the learning of this subject. It was important to her that children should learn how to do the operations, using symbols; understanding could come later. A number of the teachers, particularly those trained overseas, set the children to record addition and subtraction 'sums' in writing at the earliest opportunity. (Moreover, the work on display had clearly been directed by the teachers.) On each of my support visits the head requested discussion of these issues, but refrained from raising them when the teachers were present. Nevertheless, the staff were aware of the head's opinions and values.

Because of frequent staff absences, the head was unable to join me to see the effects of rote learning on the youngest children. In all other ways she was co-operative. She allocated £40 to each teacher for mathematics equipment. She had discussions with the key teachers after every working session. She also asked me to work with all the teachers on specific topics during the lunch breaks on the support days.

Since there was no mathematics co-ordinator at that time, I worked with each of the key teachers at every support visit. The influence of these visits can be seen in the growth of confidence on the part of one key teacher who eventually became the mathematics co-ordinator (1978). This teacher was a graduate with secondary training which did not include mathematics. She had recently returned to teaching. She taught fourth year classes for two years. At first she was anxious about her ability to control the children and taught them as a class. By the second support visit there had been a change. The children were organized in groups and were using ten-rods and units to help them to subtract. The teacher said:

> I cannot do written subtraction with these children until they understand what they are doing. It seems they have never used equipment before. They know very little and understand nothing at all.

This teacher vowed that she would continue the practical work until the children achieved understanding. She did so, and gradually gained confidence.

This key teacher and another with a fourth year class were then sharing a two-roomed hut. They exchanged ideas and gave each other mutual support. In the following term they asked the head if they could organize a games session for parents, who so often inquired what they could do to help their children with mathematics. The session proved to be most successful. This increased the confidence of both teachers and impressed the head. The working sessions had suggested activities both for the children and for the parents. They had also provided experience of group organization. It seemed unlikely that these teachers, one of whom was originally so reluctant to attempt group work, would have made these changes without the promise of continued help at the support visits and the mutual encouragement they gave each other. The teaching styles of both these teachers showed a gradual but definite change.

At the end of the school year the future co-ordinator began to have doubts again. She wondered if she had spent too much time on practical work which should have been provided at an earlier stage. (Permanent change in the attitude of this key teacher towards the teaching of mathematics came later in the year when she volunteered to take a reception class. For the first time she provided materials, observed how the children used these and framed her questions accordingly.) The other teacher, who was near retirement, began to undertake imaginative projects with the children in which she integrated several aspects of the curriculum (including mathematics).

The remainder of the staff were even more inclined than the key teachers to use formal methods. One was heard to say, 'I do not want to make any changes in my teaching of mathematics. If I do not know how to teach this subject after 20 years I should give up teaching.'

Two other teachers, the deputy head and a secondary teacher recently returned to teaching at this school, always requested help with groups of children within their classes and did their utmost to continue the activities I began. Despite the attitude of the head (and some of the teachers) and the lack of a co-ordinator, some progress was made at the support visits.

Fleet (centre-based pattern of working sessions)

The new head had just been appointed. She gave me immediate and full co-operation during the support visits. She had a special interest in mathematics and said that the project could not have come at a better time for her. The first co-ordinator (trained to teach at the secondary phase) was preoccupied with the large and difficult class she had, and was unable to concentrate on helping her colleagues with mathematics at that time. Although she said she preferred class teaching, she usually had some interesting group work in progress, but this lacked conviction and she frequently asked, 'Is this what you want?' The other two key teachers adopted an informal style of teaching. Both had clear aims in the activities they provided.

Later on, encouraged by the head, the co-ordinator studied a course in reading offered by the Open University and relinquished her post as mathematics co-ordinator. She continued to act as a key teacher. However, the head had the situation well in hand. She always had ideas about the best use of the support visits. For example, she appreciated the problems teachers had in questioning children (while they were engaged in an activity) to help them to progress without telling them exactly what to do. She therefore suggested a specific topic at each visit (e.g. volume, area, box modelling). She freed each teacher in turn, first to observe me questioning a group of the teacher's class as they worked on a practical problem, and then to take the group herself. Furthermore, the head always provided time for me to have discussion with all the staff on the day's work. It was interesting to observe how the teachers' questions became more searching as they became more confident.

There was another way in which this head helped her teachers. Before the end of the first input she organized regular staff meetings, during which a number scheme was prepared for trial in the classes. By this time all except two teachers from the former junior school were co-operating in these trials.

At the support visits both key teachers always worked in an informal way. They had various activities going on, usually based on a specific project, all reflecting the teachers' imagination. The teacher in her first post became skilled at including mathematics in all the projects she chose. (Before the second input this teacher left the borough to take up an appointment as mathematics co-ordinator at a first school in another area.)

Foster (on-site pattern of working sessions)

This school had recently moved to new buildings which were partially open-plan, after sharing the premises of Frame in cramped conditions (but surprisingly amicably). The newly appointed head, in her first headship, had not had time to weld the staff into a team before the first input. But as soon as the project began, she gave it her full and knowledgeable support.

I have rarely met so many teachers who declared immediately that they hated mathematics (including two experienced fourth year teachers and the deputy head). Perhaps this was not surprising since only 10 per cent assessed their attitude to mathematics when at school as positive, and only 30 per cent thought that their profess-ional course in mathematics had been adequate.

The co-ordinator had been appointed before the head took up office. From the outset she said that she preferred to teach older children. Although her third year class was organized in groups, it was taught as a class. She made it plain that she expected silence; she did most of the talking herself and told the children exactly what to do. She had a good knowledge of mathematics; perhaps this was why, at the working sessions, she always volunteered an answer immediately instead of giving her colleagues the opportunity to do so. This increased the difficulty she had in building up a relationship with them. Later on, the head agreed to this teacher taking a part-time advanced mathematics and science course at the local college. The course took all her spare time, so that she was unable to meet her responsibilities as mathematics co-ordinator. In the circumstances, the head suggested that she herself would give maximum support to the project.

The head's active co-operation took several forms. In her effort to improve the attitude of teachers during the working sessions she became critical of the content, and regularly offered constructive advice to make the sessions more relevant to the teachers' needs. Although I had explained the reason for the joint working sessions with the first year teachers from the middle school, and the first school head had welcomed the potential of this contact with the middle school, she could not accept that the content of the sessions should include activities for the first year children of the middle school. I understood her anxiety but I had to maintain a balance as far as the content was concerned. She requested that the sessions should take place at two-weekly intervals, to allow more time for classroom experiment between the sessions. The longer interval made a decided difference to the extent of the activities tried by the teachers, who gradually became more forthcoming about their experiments.

From the first working session it was evident that the three experienced teachers were hostile to making any change in their teaching of mathematics. Yet all three volunteered at once that they were not confident about teaching the subject. The teacher said to have the most negative attitude to mathematics announced that she had 'no particular reason for hating mathematics'. Indeed, her knowledge of the subject seemed to be above average for primary school teachers. Was the adverse attitude to mathematics of these three experienced teachers the reason for the resentment clearly shown at the working sessions? Or was it caused by the requirement that all the teachers should attend the working sessions (which were held partly in school time)? The part-time music teacher attended all the sessions although her appointment was for mornings only. The head subsequently deployed her to provide mathematical activities for groups of children in some of the larger classes.

One of the key teachers, and at least one other teacher, preferred class teaching to other methods. The second key teacher avoided teaching mathematics whenever possible, especially when I was present.

The comments made by the teachers towards the end of the first input of working sessions reflected their opinions of the shortcomings of the programme. The experienced teachers wrote:

> I don't feel I have had any basic help from the sessions. Subject matter very varied and not concentrated enough on how to introduce stages of number . . . What about 'proper' maths in books?

> Some ideas have been helpful with more able second years. Perhaps one aspect could be carried through all stages. I find individual steps difficult. I have found the sessions confusing although ideas are good.

Younger teachers, all of whom were in their first posts, also made suggestions for improving the working sessions:

> The overall programme should have been seen to be more logical and continuous I feel because my main need is for small simple steps . . . in the presentation of a concept.

> How do children learn, how much should we tell them and how much should they be able to discover?

It was evident that the teachers did not consider that the sessions had shown progression. Perhaps this was caused by the slow pace of some of the sessions and the 'interruptions' caused by the wide-ranging discussions? The two teachers from the middle school, both in their first posts, also felt the pace was slow. One wrote:

> I feel that the meetings after school were not very useful in relation to the time spent on them . . . I must admit that the after-school sessions have given a lot of useful ideas but to me these have almost been nullified by the tediousness of the sessions.

The other wrote:

> Language approach a completely new idea and rather difficult to incorporate into a classroom. Help needed with the start of activities . . . and follow-through . . . and the actual organization of group activities in a timetabled situation.

The head of the first school, who had listened to the discussion that followed the writing of the comments, wrote:

> I find the sessions revise much of the work I did many years ago and which has become part of my teaching. I find it depressing to hear my staff talk of 'knowing it all' when so little is put into practice in their classrooms.

I reminded her that I had asked the teachers to concentrate on critical comments and suggestions. One member of the group only wrote in total appreciation of the sessions. She was the head of the corresponding middle school, Missingham, who was leaving to take the headship of a first school near her home. She had asked to attend the first school sessions. She wrote:

> Most helpful and useful. Helpful in enabling me to have a starting point for mathematical activities – a base on which to build. I need to organize these ideas into activities for year-groups – a natural progression with goals clearly stated . . .

This head seemed to be the only member of this group to understand that the activities organized for teachers had to be adapted in order to be appropriate for children.

The head of Foster extended her co-operation to support visits. She always prepared a programme which included every teacher. Sometimes she asked me to work with a group of children on a specific topic, such as weighing, with all the teachers present, so that they could observe the development of the activities, the questioning and the children's responses. Between support visits the head took groups of children within each class for new activities, to help the teachers to see how to organize this work, and what questions to ask the children to help them make progress. She encouraged the teachers to try the activities introduced at the working sessions, and to allow the children to talk about what they were doing.

During the first input of the project the head introduced an innovation which had far-reaching consequences: she began to release teachers to take two children at a time for practical assessments. The teachers were tentative at first, but these opportunities made them aware, as nothing else had done, of the mathematical potential of individual children, of the value of practical activities and discussion, and of the problems some children encountered when learning mathematics. These assessments also gave the teachers experience in asking questions based on the children's reactions.

The support visits seemed to relax the teachers. They said that they had been relieved to find that I frequently had the same difficulty with children as they experienced themselves, and that there was no easy answer to helping children who found mathematics hard to learn. At the fourth support visit the head said she had already noticed that all the teachers were talking more with the children during mathematics sessions and giving them less written work.

Finally, the head arranged a parents' meeting to explain the school's policy for mathematics; all the teachers were present. The head met with no opposition from the parents.

Flanders (centre-based pattern of working sessions)

This school had shared its premises with the newly-formed middle school (Movehall) until newly adapted buildings were ready. The partnership was uneasy, mainly because of a difference of philosophy between the two heads. The head of Flanders said regretfully that her school had been described as 'traditional'. (The children's work on display was clearly teacher-directed: 'To set standards', the head explained.)

Before the project began, the head said that she did not know enough mathematics herself to offer to help the teachers. She wrote:

I have always found it easier to teach reading to children than mathematics although now I realize that the subject is vast and fascinating to study in depth. I still believe that the children need knowledge in the basic number facts and need help to achieve this rather than the 'woolly' idea that children will reach the same standard if left to find out concepts by themselves.

The co-ordinator was an imaginative teacher of all subjects except mathematics. I found it hard to accept that such a competent teacher should be so lacking in confidence in the subject for which she had been appointed co-ordinator. She equipped a spare classroom for teaching mathematics, but there her contribution ended. Later on, she wrote, 'I have tried to be of help as co-ordinator but don't want to appear bossy. Any suggestions?' When discussing her own teaching she said:

> I ought to cut back on sums . . . I feel I have become more 'formal' with experience in teaching. I like to organize and plan their work but I also like to give them chances to develop their own interests. Children get noisy with informality and need quiet working periods.

This teacher had such good relations with the children that, although much of their mathematics was taken from the blackboard or from textbooks, they enjoyed their lessons and most of them had a good knowledge of important number facts. There was a pleasant, informal atmosphere in the classroom and a hum of quiet conversation. The teacher set standards which the children respected; she was seldom deflected from her overall plan, but she rarely encouraged the children to use equipment, though this was usually available.

Support visits were seldom planned beforehand, 'because', the co-ordinator explained, 'we did not know what you wanted'. There appeared to be no coherence about the programme for these visits (despite the apparent warmth of the welcome). At one of the support visits the head volunteered that she felt that heads should have been present at the working sessions from the outset. I immediately extended an invitation to the heads of all the centre-based schools to attend the remaining two working sessions of the first input. This head came to both sessions.

At the support visits I tried to help the co-ordinator to make the most of the equipment available by using this with a group of children and developing a sequence of activities through questions. But she did not make any radical changes in her teaching style, although she said that she was providing more activities and covering less written work as the support visits continued.

The two key teachers were both in their first teaching posts and were too busy coming to terms with their own classes to think about helping their colleagues. However, they introduced some of the activities and games in their own classes. (The head said on a number of occasions that these two young teachers seemed reluctant to take advice.)

I visited other teachers on the support days – but it was rare to find that any activity I started was continued by the teachers. The head suggested at each successive visit that staff discussions should be held (with me), but these seldom took place during the first input. I was not sure that the head wanted changes to be made at that stage. Sometimes she explained that the teachers were unwilling to stay for discussion during the lunch break. Was this because of her own insecurity *vis-à-vis* mathematics? She suggested that she would like the staff to see me introducing group activities to a class, but she never arranged this. She staunchly maintained, though, that she would have attended all the working sessions herself if she had been invited.

Fowler (centre-based pattern of working sessions)

The head was one of the longest serving heads in the borough. For various reasons she had had a number of absences from school. Although she had seemed to accept, willingly, the invitation for the school to take part in the project, this attitude soon appeared to change. Objections were always raised to the dates offered for working sessions and support visits. This change may have come about when the head heard how critical the co-ordinator and one key teacher were of the working sessions.

The school had many problems. The premises were scattered and inconvenient. The new head of the middle school sharing the same site had an entirely different philosophy from that of the head of the first school, who maintained that a few of her teachers always raised objections to any changes proposed. She had found it difficult to appoint a mathematics co-ordinator. Many of the teachers had a negative attitude to the subject. Less than 30 per cent left school with a positive attitude to mathematics. Even fewer thought that their professional course in this subject was adequate. The co-ordinator's negative attitude to the subject was a serious hindrance. She gave no help to the 75 per cent of teachers who were using the new scheme in autumn, 1976. These teachers followed a part of the scheme closely, relying almost exclusively on the workbooks; only a few used the teachers' resource books, which were the core of the scheme. Some children were working well below their ability. Although the middle school used the continuation of this scheme, the co-ordinator, with a fourth year class, would not use it. 'The children must be able to do the four rules before they transfer,' she said. In discussion with the head I found that she was unaware that the new scheme was being used more than the twice a week she had requested.

In these circumstances it was not possible to give helpful support at this school. The co-ordinator introduced some activities herself during the support visits, but she maintained throughout that she was a class teacher and had too many children to provide them with practical investigations. She relied almost exclusively on a textbook.

Neither the head nor the co-ordinator encouraged other teachers to ask for help on the support days. For this reason these visits usually finished at the end of the morning, but I made a point of discussing with the head my visits to the key team. The head had asked the co-ordinator to prepare a mathematics scheme for the school, but she had not done so. Neither would she help her colleagues who had the same length of experience as she had (or longer). 'Who am I to tell these colleagues what to do in mathematics?' she had said. (At that time the LEA had not given guidance to co-ordinators.)

By the fourth support visit, the co-ordinator had announced her intention of leaving at the end of the following term. A new co-ordinator, a graduate with a special interest in language, was appointed to take up her duties when the first co-ordinator left. However, at my visits to work with groups of children during the following term, it was agreed that I should have all my discussions with the co-ordinator elect, who was to take responsibility for a new mathematics scheme, which the head regarded as an urgent priority.

Finlay (on-site pattern of working sessions)

This school, in the area of social priority (area 2), moved to new open-plan buildings at the time of reorganization. The head had been at the school for many years, first as deputy. She was often called upon to counsel parents. From the outset, the head welcomed the project.

The organization was based on team-teaching. Many of these

teachers had left their secondary schools with a negative attitude to mathematics; under ten per cent had left with a positive attitude. Just over half the staff assessed their professional course in mathematics as adequate.

The staff had decided that because their discussions had been so extensive throughout the year before the move to new premises, there was no need for schemes of work in any subject. The co-ordinator had introduced a new commercial scheme with resource books for teachers and supplementary workbooks for children. Some of the teachers were relying heavily on the workbooks. The co-ordinator did not use the new scheme with her fourth year class. She confirmed that she did little else than arithmetic, because the children would soon be transferring to the middle school. Although they sat in groups, the children were always taught as a class.

It was unfortunate that the first working session had to be a combined one for the teachers of Finlay and Measures. The afternoon of the joint meeting was hot and humid and nearly 40 teachers were crowded into a low-ceilinged 'bay'. Moreover, although this was the first session for Finlay it was the second for Measures. It was not easy to arrange a programme which would be useful to both groups of teachers.

We concentrated on situations giving rise to the four operations and the appropriate language patterns, and also on number games to help children to memorize number facts. All the teachers, but especially those from Finlay, encountered some difficulties, and consequently the pace was exceptionally slow. This did not ease relations between the staffs of the two schools. The head of the middle school was subsequently reluctant, because of the slow pace, to agree that his first year teachers should continue to attend the working sessions at the first school. Eventually, a compromise was reached, and four teachers from Measures (including the deputy head) attended the sessions at both schools.

At the second and third working sessions a determined effort was made to clarify the problems which had arisen during the first session. At the beginning of the second session the head revealed that, as a child, she had had 'nothing but fear' during mathematics lessons. 'I was in a state of utter terror,' she concluded. She also said that there had been some activities she had not understood during the first session. From then on the teachers had no compunction about intervening during any activity or game to declare that they did not understand. These reactions were at variance with their estimates of their own positive attitudes to teaching mathematics. They alternated between asking many questions and being unresponsive during the working sessions. At the fourth session, by request, an ordered sequence of activities in place value was carried out and discussed. At the fifth session the teachers showed the sequence of activities they had tried with their children and discussed problems of organization. Some maintained that they had to compromise between 'what I know to be best for the children and what the parents want in written work'.

The head took an active part in all the working sessions, encouraged the teachers to experiment with their classes, and without offering to help them herself, showed her appreciation of their efforts. It was not easy for her to realize her aims for the school, because, for an open-plan school, she had an unexpected number of teachers who were not willing to take part in team teaching. Nearly half the teachers 'preferred a quiet classroom' despite the fact that 'bays' were provided for pairs of classes to share.

The head always set a good example herself by the extensive projects she undertook, from time to time, with third and fourth year children. The mathematics inherent in these situations was exploited to the full. She welcomed the working sessions because they provided an opportunity to influence the staff in the direction of more informal teaching.

In the teachers' comments on the working sessions which follow, it was curious that so many of them referred to 'talks' or 'lectures' since, apart from discussions, they were working at activities. One teacher wrote:

I have found very useful the various activities we have practised – but have encountered organizational difficulties when using the same with the majority of the class. I would like to know an order of work for three streamed groups for, say, a class of thirty fourth years.

Two teachers from the fourth year made a joint comment:

A lot of it [the content] was not applicable to teaching with large numbers in this area.

A teacher of third years commented briefly:

> I have found the latter two sessions helpful and relevant.

She was formerly very critical during working sessions.

An experienced teacher of a reception class wrote:

> I found the first two talks were very useful and I put into use many of the ideas. Since then I personally have found things difficult to understand having not had to think of teaching maths to children over five for seventeen years, and many of the new teaching ideas are strange – difficult for me to comprehend.

The first year teachers from Measures who attended the working sessions at Finlay expressed different opinions from those already cited.

> Found the sessions interesting and informative. I have tried some ideas and intend trying more. However, I must point out that, while I've learnt many new ways I don't treat the course as 'gospel' but rather as 'refresher–plus'. I find courses of this nature interesting also as I hear other people's views and opinions.

The teacher responsible for the first two years of Measures wrote:

> I've found the sessions very stimulating from the point of view of apparatus, but I don't think enough attention was paid to the difficulties teachers face in dealing with classes of mixed ability. How does one cope with a class whose maths ability ranges from 'nothing' to 'very able'?

One of the two teachers of first year children wrote:

> I found both the theory and practical side of the lectures very helpful . . . The difficulty is in carrying out practical work (which is invaluable) in a classroom of 25 to 30 children.

It was interesting that all the teachers from Measures who attended these working sessions appeared to think that they had gained a good deal from the course, yet Finlay's teachers were more critical. Since the programme was modified from time to time in response to requests from Finlay it was unlikely that the content was in fact more appropriate for Measures.

After the initial dissension between the first and middle schools about the slow pace, regular meetings between the fourth year teachers of Finlay and the first year teachers of Measures took place. Efforts were made to ensure that there was continuity in the mathematics schemes of the two schools.

The head was always willing to discuss the progress of the support visits and to co-operate in any way she could; but she emphasized her own limitations:

> I do not have the mathematical background to offer to help the teachers in their classes.

At the first support visit I learned that the co-ordinator would be taking maternity leave before the end of the first input of the project. The co-ordinator elect was a teacher in his first post who had attended workshops directed by me during his training. He was an energetic teacher and was always prepared to introduce his own adaptation of activities or games used at the working sessions. He was well organized, used informal methods and set high standards for the children. To what extent would he be able to pass on this expertise to his colleagues, since all the teachers were timetabled to be with their classes all day?

Since the working sessions had been school-based, all the teachers at this school knew me. At subsequent support visits the co-ordinator always began by showing me the new activities he had tried with his children and the measure of his success. He then indicated which teachers would be willing to have my help. By the end of the support visits I had worked with all the teachers. Some of them, however, did not continue the activities I started or follow these up and develop them further. For example, a teacher would ask me to work with a group of children on a specific topic and would then return to her class and make no attempt to observe what I was doing. In order to encourage the teachers to make more use of appropriate vocabulary in their activities (e.g. in water 'play') I helped the children to prepare a vocabulary list – sometimes in the form of questions – which I left with each teacher for display in the appropriate section of the 'bay'.

At the next support visit the vocabulary list was not in use, and was nowhere to be seen.

Nevertheless, by the fourth support visit the head felt already that there was more talking in mathematics sessions – and less written work. However, the co-ordinator doubted, as I did, whether the activities I started were ever followed up. I enquired if he felt any responsibility for taking action about this inactivity, despite his lack of time for visiting other classes. He evaded the issue. The head promised to give the co-ordinator time to work formally with the teachers at the weekly staff meetings. She said that when he talked on an informal basis about what he was doing in mathematics, as he often did, the other teachers did not want to listen because he gave the impression that he knew all the answers.

Middle schools

Melia (centre-based pattern of working sessions)

The head was head of the former junior boys school until 1974. With the reorganization in 1974 came a number of staff changes, including the appointment of five teachers in their first posts. There were also some teachers with long experience at the school. The staff as a whole had an unusually negative attitude to mathematics. Of the 17 teachers (including the head) 65 per cent had a negative attitude to the subject at school, and 71 per cent considered their college professional course in mathematics to have been inadequate. Only 24 per cent had a positive attitude to teaching mathematics.

All the classrooms were small and the mathematics equipment was stored in the corridors (at first in locked cupboards). At my preliminary visit the head said that the equipment was not often used.

The organization of the school was unstreamed, but remedial groups were withdrawn in the first two years for reading and mathematics. In the third year, and the fourth year, the three classes were allocated to four sets for mathematics. The head, who said that until the project began there had been no staff discussion about the teaching of mathematics, immediately realized the potential value of the support visits. He offered his help in the classroom to any teacher who asked for assistance. The co-ordinator, who had attended previous courses run by me, had not, until the project began, changed her teaching style. She was a good class teacher with a satisfactory mathematical background and was willing to help her colleagues whenever she was free to do so. Of her own teaching she said: 'I should do far more practical work.'

Two young key teachers in their first posts were chosen by the head as key teachers. At the end of the first term of the project one of these left on promotion. As a replacement, the head chose another teacher in her first year. Although she had missed most of the working sessions and was not confident in her knowledge of mathematics, she was very anxious both to improve her own mathematical background and to establish the teaching of the subject on a sound foundation. Both the key teachers used the written papers distributed at the working sessions and supplemented these by additional reading.

Although three senior members of staff were resistant to change (one subsequently retired, and a second took another appointment) there were a number of teachers who were most anxious to improve their teaching of mathematics. Six of these, including the two key teachers, attended a series of workshops run by the mathematics advisory teacher; I also attended these sessions.

Even by the time of the first support visits, new activities had been introduced by the co-ordinator and the key teachers to their classes, and working groups were well established. The new key teacher, who had attended only two of the working sessions, and who had been most unsure of her ability to teach mathematics, said that she had learned all she knew about this subject 'on the job', from her colleagues. She had had particular help from another key teacher – once more, an example of the benefit of mutual support. This probationary teacher had her second year class organized in groups according to ability, each group working on a different practical problem. She said that she now used textbooks only for providing 'practice': she had become independent of them.

By the third visit the head and the co-ordinator said, 'The project has lifted mathematics off the ground. This could not have happened but for the project.' The co-ordinator had already had meetings with the teachers, particularly with the key team. She had also given support to those teachers who asked for help in their classrooms. Moreover, some teachers had paid reciprocal visits to the co-ordinator's classroom.

The head made full use of the support visits in every possible way.

During the lunch breaks there were always groups of two or three teachers who asked me for help with some specific aspect of mathematics. Sometimes short workshops were organized on selected topics (such as the introduction of fractions and decimals, place value). In general, the reactions of the teachers and the head at this school to the support visits were positive and changes ensued in the teaching of mathematics. The key teacher who had been most insecure about the teaching of mathematics said: 'I enjoy teaching maths so much now that I should like to do it all day long.' The head said: 'I could not have helped teachers to improve their teaching of mathematics if it had not been for the project.'

This promising beginning was not developed further by that co-ordinator, who left at the end of the following term on maternity leave. The co-ordinator elect was an experienced teacher recently appointed to the school. She did not have a strong mathematical background, but she was willing to remedy this and she looked forward to her new responsibility.

Meakins (centre-based pattern of working sessions)

The head took up this, her first headship, in September, 1974. Her teaching experience had been at the secondary phase and she had attended a previous mathematics workshop I had run in the borough. She was knowledgeable in mathematics and supportive of the project, but she did not take an active part in teaching any subject on a regular basis. There were a few very experienced teachers at this school. Four new appointments had been made in 1974, all of teachers in their first posts. Of the five appointments made in the following year, three were in their first posts.

The classes were unstreamed and there was no setting for mathematics until the fourth year. The co-ordinator was particularly well qualified in mathematics and science, for both of which he had responsibility. He attended all the mathematics courses subsequently provided at the teachers' centre by the advisers. As a teacher he presented his material in an imaginative way. He was successful in establishing good relationships with children of all abilities. He showed a preference for class teaching, which formed part of every lesson, but the children were then allowed to work in groups or as individuals as they pleased. He vacillated about the effectiveness of working in groups.

From the beginning of the project he was hesitant about helping his colleagues, particularly those who were experienced teachers. The head gave him all the support she could, without taking an active part in the teaching herself. She organized staff meetings at which he provided games and other activities for the teachers; he also organized a 'sponsored test' of number facts for all the children in the school. The head allocated some non-teaching time for him to enable him to work with his colleagues on request. However, he chose to use this time to work with small groups of able children whom he withdrew from first year classes. From his point of view this was a useful exercise, since he was able to find the extent of the children's understanding of concepts – and also their mathematical potential – but he did not pass this information to the teachers concerned.

The teachers were expected by the head to include mathematics in the projects which were a feature of the school, but there was no evidence of this subject in the displays. The staff were given no help in this respect by the co-ordinator.

One lively key teacher was willing to try any activity with her children; she was skilful at adapting these to suit the needs of children of different abilities. She organised her class successfully in groups. Her enthusiasm infected other teachers in her year-group and she always discussed the papers from the working sessions with them. She took a lead as a key teacher from the beginning of the project. The other key teacher was anxious about class control when trying activities; she was slow to gain confidence in teaching mathematics (in which she had had no training). She was more confident with a younger class but she was happiest when she was using a textbook or workcards, although she continued to use some of the activities suggested at the working sessions.

The experienced teachers at this school followed textbooks closely. Those who were using the workcards did so as recommended, on an individual basis. There was little group or class teaching and there were few opportunities for discussion during mathematics lessons, except in the second year (led by the key teacher who was keen to experiment).

At the support visits I made myself available during the lunch break for staff discussion, but few teachers made use of this opportunity.

The first input of the project was not as effective as might have been expected at a school with a knowledgeable and supportive head, and a co-ordinator with an unusually good mathematical background.

Missingham (on-site pattern of working sessions)

Setbacks at this school were caused by:

1. the staff turnover at senior level;
2. the number of young teachers in their first posts — six of the total of thirteen teachers;
3. the fact that two teachers trained for the secondary phase had had no professional course in mathematics;
4. the absence of a policy or of any coherent scheme for mathematics; there were many different textbooks and workcards in use in different parts of the school;
5. the low assessments by the teachers of the adequacy of their professional courses in mathematics; 62 per cent thought them inadequate.

Nevertheless, at that time the setting for mathematics throughout the school (two classes into three sets) at least resulted in smaller teaching units.

The former deputy had leave of absence during the first term of the project. She returned as acting head and was later appointed head. Although the acting head had a good mathematical background and a positive attitude to the subject, she was concerned with far too many problems of a general nature to volunteer to help the teachers with mathematics herself. However, she kept an eye on young teachers who were making changes in their teaching styles and was ready with help and encouragement when they ran into difficulties. She warmly welcomed me and was frank about the problems the teachers experienced.

The three key teachers were chosen by the former head (who left at the end of the first term of the project). Two in their first posts were coming to grips with their individual teaching problems; the third, who had a strong mathematical background, had organizational difficulties of her own. None was ready to act as a key teacher at that time. In view of these circumstances it was not surprising that the teachers were not always responsive during the working sessions.

Perhaps because of the presence of a mathematics teacher from the high school, or because there were so many teachers in their first posts who did not yet feel secure, particularly when teaching mathematics, they rarely volunteered that they did not understand an activity or a game. Neither did they raise any objections to the introduction of more than one method of carrying out a calculation. On the contrary, some of them immediately tried the different methods with their children. Yet the teachers with more experience had a marked preference for class teaching.

It was encouraging to find that so many young teachers experimented in their classrooms between the working sessions, and were forthcoming about their failures as well as their successes. By the final session the group was relaxed and showed a ready sense of humour. But since there was no co-ordinator to appreciate their efforts and to help them with their problems, it was not surprising that these efforts were not sustained.

The following written assessments were made at the fifth working session after one support visit. The acting deputy wrote:

> Interesting and enlightening. Division – very confusing – would like more guidance . . . I need an overall picture or scheme of work to know where I am going – to know how everything links up – how to progress from one aspect of mathematics to another. This stems from a lack of (a) knowledge and (b) confidence re. mathematics and teaching of.

It was a great loss to the school that this candid and co-operative teacher left when he was appointed head of another school before the end of the first input.

An experienced key teacher of fourth years wrote:

> Found it very interesting from own point of view. Think I will find it helpful in future – particularly measuring activities – with older children. Course has made me realize the *difficulties* children have – with concepts and language. Used to get impatient when they couldn't understand . . . I never found it difficult when young – beginning to realize how difficult it can be *now*.

The remaining comments were made by teachers in their first or second year of teaching. A key teacher of a first year class wrote:

Some of the work (e.g. subtraction) I have found interesting and useful. Many things I have not been able to organize . . . owing to the size of class. Some of the activities I have found tiresome myself – cutting up paper, measuring, etc. – especially after a day at school.

(During the working sessions this teacher had given no hint of her attitude to some of the activities.)

Three other teachers in their first or second year of teaching also gave their opinions. A teacher of fourth years, who clearly agreed with the preceding teacher, wrote:

I found activities for use in the classroom useful – sometimes we spent too much time doing things which were not intellectually stimulating – leading to boredom.

A teacher of third year children wrote:

Helped very much in providing an even better attitude towards teaching maths – and provoked even deeper thought about method of teaching the subject.

A teacher of second years wrote:

These sessions have helped me to put new ideas into practice as well as develop old ideas.

Despite the many recent appointments, particularly of teachers in their first posts, there was potential for change. But without a co-ordinator to encourage the teachers, to give them advice when necessary and to spur them on, they were at a disadvantage.

As soon as the support visits began, they aroused the interest of several members of staff who made strenuous efforts to introduce a variety of activities and games to the children they taught. I was able to work with nearly all the teachers at each of the support visits. The key teachers and some of the new teachers who were trying group activities with difficult classes asked for support at every visit. Problems of control were sometimes aggravated because the teachers preferred to work in their cramped classrooms, rather than the spacious, but dark, hall allocated to mathematics teaching. However, progress was made in the control of working groups as well as in the coherence and development of the activities used.

One young teacher made a remarkable change in her teaching of mathematics over a short period. At my initial observation visit she was using a workcard system with her class. Much time was wasted because of the queues of children, waiting either to ask questions or to have their work marked. But before the end of the first term of the project this teacher had decided not to use the workcards again. She explained, 'I cannot put enough of myself into the teaching with these cards.' During the second term of the project she asked me to introduce her second year class to long division. I found this session difficult to organize because I did not know how much understanding these children had of the concept of division. However, the teacher wrote in her first assessment of the project: 'I was surprised at how much I enjoyed and learned from the support days.' From then on she pressed for longer discussion periods before and after each support session. One day she asked me to help her with the planning of her project work so that she could include mathematics. She began the discussion by saying 'Don't talk until I have told you where I need your help. I know exactly where this is.'

The ensuing discussion was terse and to the point. Moreover, this teacher also influenced a young colleague with a particularly negative attitude to teaching mathematics, with whom she worked in partnership. With her support he changed his class teaching to an organization which facilitated group work whenever mathematical activities were in progress. He gained confidence when teaching mathematics through working with this teacher. In this way she fulfilled her responsibilities as a key teacher.

Some other teachers, including five of the six who were in their first posts, began to change their teaching style for mathematics. They made determined efforts to plan sequences of activities and to encourage discussion. This may well have developed because the working sessions were school-based and therefore all the teachers knew me and were familiar with the activities themselves.

Movehall (centre-based pattern of working sessions)

Preliminary visits were made before the school moved to a new site. The head was in his first headship and five of the ten teachers were in their first posts. He began to put his ideas into practice immediately by introducing vertical grouping for the first two years and team-teaching. There was an experienced deputy and a senior woman who was also mathematics co-ordinator. The head told me that, although most of the work in the school was integrated, mathematics was taught as a separate subject.

The co-ordinator had made an outstanding contribution at the previous workshops in mathematics and had helped to prepare the LEA guidelines in this subject. She wrote: 'I rarely use a textbook because this takes me away from the children'. She was a teacher of rare quality who planned her work on original topics, but based the development entirely on the children's responses to her skilful questions. Every child took an active part, not only in the activities but in all discussions. This teacher used informal methods most successfully.

The school was labouring under many difficulties at this time, mainly because the head and the teachers were preoccupied with the impending move to new premises. I realized that during the support visits (which began just after the move) I could not expect active co-operation from the head in terms of teaching himself or offering to help the teachers when they were experimenting with their own classes. Furthermore, the co-ordinator had many other responsibilities.

There was no special scheme for mathematics at that time; the borough guidelines were in current use. In the new premises there were two spacious rooms for mathematics, well equipped by the co-ordinator. Because of setting for mathematics the teaching groups were small. This offset the short weekly time allowance of 2½ to 3 hours – but the organization of this time allocation into two or at most three lessons led to some lengthy mathematics periods (1¼ to 1½ hours). Teachers with little experience found that they had to plan carefully in order to retain the children's interest and sustain a good pace of work.

It was unfortunate that the teaching skills of the co-ordinator, who had already made her mark in the borough, were not used as an example for her less experienced colleagues. (It was a great loss to the school when this co-ordinator left in December, 1976, at the end of the first input of the project, on maternity leave.) The two key teachers were in their first and second years of teaching. Eventually these two became joint co-ordinators for mathematics, but they were not ready to take this responsibility when the post became vacant. Although both were promising teachers they were relying extensively on textbooks at that time. Already, by the fourth support visit, one key teacher had come to grips with her major problem, the needs of slow children; the other had prepared a varied collection of games and activities for the children she taught.

At all the support visits I was able to work with the teachers concerned with mathematics. Most of the activities I introduced on request were followed up by the teachers. The children were encouraged to make attractive displays of their work in the corridors. These displays led to discussion among other children and were instrumental in leading to further changes in the teaching of mathematics.

Two young teachers with little confidence in their own ability to teach mathematics asked for special help. They were uncertain because of their own mathematical backgrounds. They received much encouragement from both the co-ordinator and myself. One of them was particularly reluctant to undertake investigations (on scale) because her group of able children had already 'worked all the examples on scale from the textbook'. She concluded that the children understood the concept and thought that the investigations were unnecessary. I suggested that these could be used to assess the extent of the children's understanding. The young teacher (an Arts graduate) still hesitated; later on, she confessed that she had doubted her ability to describe the investigation to the children and help them to carry it out. When, finally, she was urged yet again by the co-ordinator to try the investigation with the children, the teacher was surprised both at the children's enthusiasm for the investigation (making full-scale, half-scale and quarter-scale three-dimensional models of themselves) and at their lack of understanding of the concept of scale. (Moreover, from then on, the children assessed mathematics as their favourite subject.) This young teacher, in her first post, had acquired a good knowledge of mathematics at school, but had assessed her professional course as totally inadequate. She was a successful teacher in other aspects of the curriculum, yet she asked for further help. 'How shall I develop this topic next? I think they need further real experience,' she said. I arranged for a lengthy discussion to enable this teacher to find resource material for herself and to help her to become more confident.

This experience gave me an insight into one of the greatest

hindrances to sustained innovation in mathematics: the failure to provide adequate support by making further resources available. This competent young teacher was working with only nine able children (for mathematics), so that there was no difficulty with control, yet still she had hesitated. She said that if she had not been encouraged by the co-ordinator, for whom she had a great respect, she could not have overcome her fear. The success of this activity had changed the attitude to mathematics of both the teachers concerned. They continued to develop their mathematics teaching, basing this on practical investigations and asking advice from time to time from the co-ordinator or from me.

The head was welcoming, and discussions with all the teachers were arranged whenever possible. The support visits seemed to have created a favourable climate for the project, perhaps because the philosophy of the head encouraged informal methods of teaching. It was therefore not difficult to effect changes in the teaching styles for mathematics of four young teachers. Initially, the most important factor in these changes was the continued support and encouragement given by the mathematics co-ordinator. Later on, the support visits became the motivating force.

Makewell (centre-based pattern of working sessions)

There had been fifteen changes of staff during the two years since reorganization. Several of the new teachers were in their first posts and the head, appointed in 1975, was in his first headship. Because of his interest in mathematics, which was well known in the borough, he had been reluctant to make changes in this subject before he had appointed a co-ordinator (from the school at which he had been deputy head). He therefore appointed four key teachers, all of whom attended the first input of the project. Two of the key teachers were experienced; the other two were in their first posts and in their second year of teaching.

There was setting according to mathematical ability in the third and fourth years; four classes were allocated to five sets. The teachers at this stage benefitted from the smaller numbers in each set – but the numbers in all the classes were large at that time.

The attitude of many of the teachers to mathematics was negative. Although only 20 per cent of the 15 teachers who assessed their attitudes to mathematics said that they were not confident in teaching the subject, an unusually high percentage (about 50 per cent) assessed their attitudes when they left school or college as negative. The head had described the mathematics at the school:

> Maths is formal through the school. Most teachers are determined that children shall learn the four rules before they learn anything more exciting.

The co-ordinator was appointed one term after the project began. The head and this co-ordinator immediately began to discuss the introduction of an up-to-date (commercial) mathematics scheme, based largely on source-books for teachers. Both were determined that this scheme should be used as intended, and soundly based on the practical activities and games suggested by the source-books. To this end, the head arranged two discussions with the teachers to launch the new scheme. To help them to organize practical investigations with groups of children (and so to persuade them of the importance of peer discussion) the head offered to reorganize those classrooms in which the desks were arranged in rows and to make spaces for reading and mathematics 'corners'. The co-ordinator assisted the head in this venture.

Although the co-ordinator was the senior woman, and therefore had other responsibilities, she regarded herself first and foremost as the mathematics co-ordinator, and the head gave her every backing. He released her to work with her colleagues in their classrooms. She had discussions with them in year-groups, often at her home. She was soon able to assess the strengths of the mathematics teaching and where help was needed. She realized that some teachers, even those with experience, who felt insecure when teaching mathematics, might be intimidated by her position, and arranged that those teachers who were confident and skilled at teaching the subject should help others who lacked confidence and asked for help. In this way she gave some of her colleagues valuable experience in working as 'key' teachers. She planned the support visits to best advantage and always made it possible for me to work with individual teachers, or small groups, in the lunch break.

Another way in which the head and the co-ordinator joined forces was to conduct both discussions for school managers and workshops for parents to inform them of the school's policy concerning

mathematics. Furthermore, visits were arranged to the corresponding first school and to the high school, to establish contact and to try
to ensure continuity in mathematics teaching – in content and in
method. All this was achieved before the end of the first input.

In view of the encouragement and practical help given within the
school to the key teachers, it was not surprising that three of the team
of four changed their teaching styles to facilitate the use of practical
investigations with groups of children by the time of the second
support visit. (The least experienced teacher continued to persevere
despite the difficulties she had in controlling the children.) One, a key
teacher with six years' experience, lacked confidence when teaching
mathematics and assessed his style as 'formal'. He wrote:

> I am formal, in that my maths set do the same work at the same
> time, at varying speeds. They are not allowed to move freely
> around the classroom, as this inevitably leads to friction, if not
> physical violence. My attitude is informal, in that children can talk
> about and discuss their work at will, and call me over when I'm
> required.

The head also described this teacher as 'a formal class teacher'. This
teacher asked particularly for help with practical work, saying that
he used textbooks too much. Yet between two support visits his
teaching style changed from formal class teaching to well-organized
groups, in which pupils were discussing their work and were allowed
to select the materials they used. This, of course, necessitated moving
around the classroom. When I discussed this rapid change with the
co-ordinator, she said:

> [His] change of style was influential because he was experienced
> and had never before shown any inclination to change. Although
> he did not try to change other colleagues the changes he made were
> so evident that others who had set their faces against change were
> influenced.

By his own example this teacher made his contribution as a key
teacher. When questioned by the advisory teacher about his work
this teacher answered:

> I would not have changed but for the course. I have now totally
> changed my attitude – I have gained immensely. At first I was anti

— then after trying activities at the working sessions, I decided to try using materials with my pupils — for the first time since my teaching practices, where this was expected. I was anxious lest I could not control the pupils in this situation. I began by giving the children the same thing to do — area of hands. I found the children finished at different rates so that I was able to start them off on the next stage. I found that I could control this situation easily. I now let the children work at different activities at the same time.

It was unusual for a teacher, even an experienced one, to make such a rapid change in his teaching style, the more so in view of his confessed hatred of mathematics at my first observation visit. He was a teacher of physical education who had no difficulty in controlling his classes in that subject. (Before the second input he left to become deputy head at another middle school.)

A second teacher made his decision about the optimum method for him, of teaching mathematics, at the first observation visit. He chose to organize practical work in groups; he attributed this choice to the children's response, to the ease with which discussion followed, and to the encouragement he received from me at the time. At the fourth working session he assessed his teaching as:

Formal in that I decide what each group is to do and expect them to listen to directions and questions at the appropriate times. Informal in that the children work in groups, help one another and discuss problems among themselves and with me. I do expect quiet from the groups if I am working with a particular group. I do not encourage wandering round the room unless the nature of the work demands it.

At his first interview this teacher had asserted his satisfaction with his professional course in mathematics. At the first observation visit he had announced his intention of making that lesson 'a one-off practical lesson'. In his assessment of the working sessions he wrote that he had been 'rather daunted by the amount of practical work needed', and he found it difficult 'to know what it would be valid to use in teaching children of ten'. By the time he left the school to become a mathematics co-ordinator at another middle school, his teaching had gone from strength to strength. He had never had difficulty in controlling his classes. By then he was fully confident in his ability to plan successfully and put into practice a mathematics programme for any level from eight to twelve years old. (He subsequently attended a two-year part-time Mathematics Diploma course for teachers.)

I also worked with several other teachers suggested by the co-ordinator. One or two were resistant to change, particularly a former high school teacher. But the head and the mathematics co-ordinator formed a strong team with a common purpose and a number of changes were initiated during the first input of the project. Both the head and the co-ordinator helped new teachers in their classrooms to improve their teaching of mathematics. They also encouraged others to increase their mathematical background by reading or by attending courses. Four took advantage of this opportunity.

Measures (school-based pattern of working sessions)

This school was in an area of social priority (area 2), and was well established in the area. Reorganization had not been accompanied by a high staff turnover; few new appointments had been made. The head had found a new textbook in use throughout the school when he was appointed ten years previously. This was still in use, supplemented by more recent books and by a system of workcards introduced in the first year classes. There was setting for mathematics through the school.

The head had attended a national primary mathematics course I had organized more than 15 years previously. He had retained his interest in mathematics and said that the course had given him confidence to question any class on the subject. From the outset he was most co-operative. The co-ordinator had attended the previous mathematics workshops in the borough and had been one of the team responsible for preparing the borough guidelines. By the time of the first input she had already organized and distributed the equipment, and she was preparing a scheme. Unfortunately, as senior woman on the staff she had other calls on her time. At the observation visits she showed herself to be a competent 'formal' teacher, asking searching questions, setting high standards of achievement and expecting good standards in the presentation of written work. But it was evident that she was not using the investigations included at the original workshops, or providing opportunities for discussion. She was one of the few teachers who had always had a positive attitude to mathematics.

At the observation visits the school gave the impression of being formal, perhaps because the head thought that the children in this difficult area should work in a quiet atmosphere; his staff supported him. Two of the key teachers seemed to be more formal than informal with their classes. The other two had more informal contact with the children which offset some of the formal tendencies of the teaching.

The head was always present at the working sessions. The three mathematics teachers in the first year of the high school were also present. There were useful exchanges, for the first time, between the teachers from the two schools. From the beginning, the teachers were relaxed at the working sessions and always ready to ask for further explanation when they needed this. The sessions were characterized by critical questioning and more discussions than at the other schools with the on-site pattern. Most of the teachers tried various games and activities between the sessions and were prepared to discuss their experiments subsequently. The head and his deputy always remained after each session to discuss the problems the teachers had raised and to appraise the sessions.

The head encouraged the teachers in their experiments, particularly in their provision of practical activities followed by discussion among groups of children. He emphasized the value of this method for assessing children's understanding. By contrast, the deputy, who always taught very able children, consistently maintained that he preferred class teaching, although on this basis he was happy to introduce a number of the activities. It was interesting that both the head and the deputy assessed their own teaching as 'formal'. The head wrote that the project had been most useful:

> . . . the importance of correct attitudes to encourage children . . . Flexibility in approach to mathematics . . . In time allowed, topics discussed have been very important as fundamental.

The co-ordinator wrote:

> Yes it has been helpful in that it has made me take another look at my teaching methods. Some very useful ideas, but sometimes we tended to rush from one thing to another which made it a little confusing. In future sessions it would be useful to take specific

topics and follow them through, showing how to begin a new topic and how to carry it through to the harder stages.

This request was met by the planning of a sequence on place value; this was extended to decimal fractions, scale and graphical representation, through a series of activities.

The senior mistress, who had had a consistently negative attitude to mathematics, wrote:

> Stimulating and thought-provoking, but still leaving me a bit at sea with the underlying organization and direction of maths, which covers so large an area. I have appreciated many useful and helpful new looks.

Another experienced teacher wrote:

> Helpful. I am reassured and spend more time on games and activities. Need more help in teaching tables e.g. games or helpful activities.

An experienced teacher stated one way in which the sessions had been helpful ('thought-provoking ideas') and one way in which the sessions were unhelpful ('I would hesitate to introduce different methods to children of lower ability – which may point to my teaching inadequacy!'). A third experienced teacher was also critical:

> On the whole the lectures have been useful, but occasionally I have felt that my time has been wasted because so much attention has been spent on teaching concepts which are too basic for my particular maths set . . . I found the suggested methods of teaching division, multiplication and fractions particularly helpful.

Some teachers in their first posts had found the sessions more useful. Three of them wrote:

> Yes very helpful – ideas for practical work especially for less able sets. Has also influenced my attitude to maths – built more confidence and a less tense and rigid manner when teaching maths. More creative work possible in maths than I had thought.

Helpful in that it stimulated thinking in mathematical terms, though it did not necessarily break new ground.

Interesting from a personal point of view. Entertaining and a useful 'reminder' i.e. I feel that periodic instruction through doing maths oneself, helps to revise otherwise possible 'stale' ideas.

A teacher in her second year wrote:

These sessions have been helpful to me. They have cleared up points of confusion in my mind. I have found particularly that I can now explain things more clearly to the children. However, I still have not tried many of the things with my group as I am not always sure how to apply them in the classroom situation. Also we often cover so much that I do not retain much of it.

The last account was particularly interesting in view of the co-ordinator's comments on the confusion she experienced by moving too quickly from one topic to another. This often resulted from the questions teachers asked, and their responses to the activities provided. I have already described the determined efforts made in subsequent sessions to provide sequences of activities on topics chosen by the teachers themselves.

From the similarity of their comments the two high school teachers had evidently discussed the sessions before they wrote their comments:

The sessions have been helpful in clarifying certain problems, like the knowledge of number facts – the activities approach I find attractive but I have not yet been able to assess its effectiveness in terms of ordered results in a classroom (20 to 30 children) situation.

A third young high school teacher wrote:

Helpful in that the course has indicated a new/fresh approach which I had not previously come across.

The comments made by the teachers at Measures were more positive than those made by the teachers at other on-site schools. This may have been because the head himself was confident about teaching mathematics and was always ready to give encouragement to those teachers who were willing to experiment. The comments were in tune with the receptive atmosphere of the working sessions and the pleasant give-and-take which characterized the sessions.

At the support visits the co-ordinator gradually relaxed and provided the children with more opportunities for discussion and for undertaking practical activities. She was always prepared to encourage her colleagues and to discuss the problems they met when trying to change their teaching methods in mathematics, but since the school occupied two different buildings, not far apart, she did not have frequent contact with the teachers in the lower school.

The head's knowledge of the strengths and weaknesses of his staff was invaluable. He always tried to be present at my discussions with the co-ordinator. Many of these discussions centred on the dichotomy between practical activities, devised to help children to understand concepts (such as the four operations) and written calculations which were usually set from books. All too often, there was no transitional stage and teachers and children failed to connect the two processes. Although the head did not offer to help the teachers to implement the activities introduced at the working sessions in the second half of the first input he was able to experiment himself when he taught the lowest set in the third year on a regular basis. He gave them activities to assess their understanding of concepts, and provided plenty of opportunities for discussion. He frequently described the children's responses to the staff and this encouraged the teachers in their own experiments.

The support visits were well planned by the co-ordinator and the head. Many of the teachers asked for help at each visit. Five of them, who hitherto had always taught mathematics on a class basis, working from textbooks, and providing little opportunity for investigations, asked me to help them to organize group activities. Some of these teachers had set their pupils to complete exercises from the textbook, for example on fractions, and were surprised to find that practical activities revealed how little the children had understood what they were doing. Not all were convinced that activities would help children's understanding, or that their discussion could show whether they were ready for textbook practice or not. One young teacher persisted in saying that children should learn by rote – understanding was not necessary.

Time was always made for appraisal of the lessons and for

discussion of the future development of the concepts considered. One of the most valuable features of the support visits was the informal discussion that I was able to have in the staff rooms. Queries about the optimum organization when using activities were often raised by individual teachers who were too shy to do so at the working sessions. I also noticed how many teachers gained in confidence during the support visits. (This applied to the co-ordinator, too; the head's confidence in her judgements also increased.)

Although neither the head nor the co-ordinator was able to help and encourage teachers in their classrooms, some teachers, including some of the most formal, gained enough confidence to provide practical activities for groups of children. Because of the tradition of quiet classrooms the problem of controlling the class during such sessions was not as evident as in some schools. But perhaps this made some of the younger teachers more apprehensive about experimenting in this way.

An interim comparison of the effects of the school-based and centre-based working sessions

The major differences between the effects of the working sessions for individual schools and those held for the nine other schools at the teachers' centre were caused by the presence of the head and the entire staff at school-based sessions and the fact that the teachers at these schools, with few exceptions, were known to each other. It was at once apparent, at every school-based session, that the presence of the head made a great difference in some ways.

1. The working sessions were an unobtrusive way of informing the heads, at first hand, of the aims and content of the project. At least one of the heads learned a good deal of mathematics she had not known before.

2. The heads were therefore given confidence themselves, and were more likely to follow up points made during the sessions and to encourage the teachers when they were experimenting.

3. At the school-based sessions the content could be 'tailored' to the expressed needs of the teachers.

4. The atmosphere of the sessions was more relaxed. At the teachers' centre, teachers were usually somewhat preoccupied with getting to know all the teachers from the other schools, particularly those to whom they would send, or from whom they would receive, children. At school-based sessions there were fewer teachers unknown to the group and contacts were established at an early stage.

5. The teachers at these schools came to know each other better; the sessions were of special value to the co-ordinator who was able, during the working sessions, to assess the strengths and weaknesses of all her colleagues.

6. I came to know all the individual teachers and particularly their attitudes to mathematics; since only the teams of co-ordinator and key teachers were present at the teachers' centre I was unable to get to know the remainder of the teachers at those schools until the support visits.

It therefore seemed, at that time, that the schools having on-site working sessions were definitely in an advantageous position compared with centre-based schools.

Interim conclusions after the first input

Before the first input I assessed that, of the 39 co-ordinators and key teachers, less than 20 per cent were using informal teaching methods for mathematics (including activities and discussion). By the end of the first input, three heads, two co-ordinators and eight teachers (seven of them key teachers) had made radical changes in the teaching of mathematics.

The three heads (of Fleet, Foster and Makewell) rapidly took advantage of the project to effect changes in their schools. Because their methods were individual and therefore different, I wondered whether they had any characteristics in common. All three were in their first headships and near the beginning of their careers. Each was confident in her knowledge of mathematics and in her competence to teach the subject. Each was determined to introduce some informality into the teaching (there were some formal teachers at each of the schools), but in due course, when the teachers had been prepared for this change.

The speed and extent of the changes made were surprising in view of the fact that the conferences organized by LEA advisers to inform the co-ordinators about what was expected of them were not held until the end of 1976 and the beginning of 1977. In consequence co-ordinators, particularly those at first schools, were bewildered about what they should do and worried about visiting the classrooms of their colleagues, either for assessment or to give help. None of the first school co-ordinators had yet begun to act as leaders. All three of these co-ordinators, who later became operative, claimed to have a most inadequate knowledge of mathematics. This was not true of the middle school co-ordinators: two fully effective co-ordinators (at Movehall and Makewell) made an immediate effect on their colleagues' teaching by their own good examples. Two others (at Melia and Measures), both of whom had attended a previous course I had run, gradually changed their teaching of mathematics. There had been evidence at the original follow-up course that these teachers had begun to make changes at that time, but these had apparently ceased after a while. What had brought about the more recent changes? Were these the result of the support visits? Or were the working sessions at last having an effect? (All had an adequate knowledge of mathematics.) Or was it because the co-ordinators realized that if they were to influence their colleagues their own teaching must exemplify the changes they planned to achieve? Whatever the cause, the middle school co-ordinators quickly came to terms with their responsibility for assessing the strengths and weaknesses of those who taught mathematics. The support visits came at an opportune time for them, to show them what was involved in helping individual teachers in their own classrooms, and the potential of such help. Furthermore, the support visits prepared the teachers for similar visits from the co-ordinators. All except one of them were prepared to assist young teachers and other colleagues who asked for help. They were all hesitant about giving advice to more senior colleagues.

With two exceptions, the eight teachers who made rapid changes in their teaching of mathematics were in their first posts and in the first three years of their appointments. Only two were at first schools. All of them made an indirect contribution to change within their schools by their own examples. They successfully adapted their teaching styles, adopting a group organization for the activities they introduced and encouraging discussion. This was not an easy innovation for inexperienced teachers to achieve. All except one (who left on maternity leave) ultimately received promotion when they took up other appointments.

In addition to these 'early adopters' there were a number of heads, co-ordinators and key teachers who had begun to experiment but proceeded at a slower pace. Had it not been for the regular support visits, these teachers (and possibly others unknown to me) might well have given up the struggle. I knew of two heads, two co-ordinators and sixteen teachers who belonged to this group.

There were, of course, early resisters to change (including two heads). A few did not agree with the underlying philosophy of the project. Some were diffident, others were set in their ways, and many felt that their knowledge of mathematics was inadequate. Three of the co-ordinators were diffident and therefore were unable to help their colleagues. Some of these teachers improved their mathematical knowledge by working through the assignments of a new workcard system, others attended additional mathematics courses run by the advisers. But the most recalcitrant of the resisters were those, often with more than five years' experience, who readily agreed to try some activities, but did nothing.

There was a further incentive to change. Although some key teachers were not sufficiently confident, either to give their colleagues an account of the working sessions or to impart their growing enthusiasm, they gave each other support in their experiments. This increased their confidence and often led to renewed efforts with their children.

In conclusion, it was evident that only a beginning had been made in the utilization of the support visits at three of the first schools. Once these schools had taken part in the LEA conferences on the function of co-ordinators, I hoped that these teachers would begin to assume their responsibilities. They might then make better use of the support visits, as the co-ordinators at the middle schools were already doing. The working sessions played an important part in initiating the desire to change by providing activities and games which teachers could do, could enjoy and could attempt in their own classrooms. But if teachers became frustrated by problems of organization, more help was required. It was at this stage, when teachers needed assistance with their classes, that the value of the support visits became apparent.

CHAPTER 6

The second input and an assessment of the project to date

Preparation

Three factors influenced my plan for the second input: the teachers' reactions to the first input, described in Chapter 5; what I learned by working with groups of able and slow-learning children in all the project schools; and the second round of interviews with the heads and selected teachers.

The objectives and outcomes of nine sessions with groups of children

These sessions took place during the spring and summer terms of 1977. With the intention of seeing what the teachers in project schools, particularly the co-ordinators and key teachers, could achieve without my active support, I kept out of their classrooms. At the same time, the sessions enabled me to maintain informal contact with the schools; I was available for consultation and kept in touch with developments and the problems arising. Since I planned to discuss with the teachers the progress of the children in the groups, I hoped that the teachers would realize that I was experiencing, at first hand, some of their teaching problems.

The slow-learning children were in groups of four to eight, and the able children in groups of six to nine; each group (chosen by the head and the teachers) had a two-year age range. The sessions occurred at weekly or two-weekly intervals. As much ground as possible was covered by informal methods: activities and discussion. Although I prepared a programme, I utilized suggestions from the children to the fullest extent, and I provided attractive materials throughout, both for problem-solving and for the presentation of results. The sessions were taped when this was practicable.

Details of my work with the children are set out elsewhere. For the purpose of describing the influence of the sessions on the second input, the following account of my objectives and the extent to which these were met is sufficient.

1. To get to know some of the children in each project school, particularly those who created problems for their teachers: the able, whom it was hard to keep fully occupied, and the slow who found it difficult either to understand concepts or to learn number facts.

2. To identify the special learning problems of slow children and to try to find ways of overcoming these; to provide able children with a variety of investigations and to note their reactions. To try to discover the extent of children's knowledge of number facts and their understanding of concepts and calculations. In subsequent discussion with their teachers, the content and teaching methods used would be described; the teachers' own understanding and knowledge of mathematics might also be revealed.

3. To observe the attitude of individual children to mathematics and to notice whether there was any change over the course of the sessions.

4. To encourage some of the teachers to continue the activities I started.

All these objectives were met to the following extent.

1. I formed a clear impression of each child's strengths and where he or she needed assistance. Many teachers expressed appreciation during my subsequent discussions with them; it encouraged them to find that I experienced the same difficulties as they had with individual children. Frequently, the teachers supplied background knowledge of individuals which helped me to deal with their problems successfully.

2. Few of the children were knowledgeable about the different situations which require the use of one of the four operations for their solution; neither did they know the appropriate language patterns. Establishing these patterns was a long process for some children, and a routine needed to be put into operation. On the other hand, able children could tackle all the investigations I provided, and often responded enthusiastically. A few of the able children, especially the older ones, and occasionally some of the slow-learning children, were scornful about the material and equipment I always provided. A 12-year-old boy described this as 'childish'. I had explained that the children could always use their imagination if they preferred to do so. Ultimately they all made use of the material at one time or another, especially when they found that other children were quicker at discovering a solution when they used it. It was valuable to be able to describe to the teachers various problems which the children had been able to solve because appropriate material was available. Some teachers had formerly been reluctant to provide material.

In general, many children used incorrect terms when referring to people and objects; for instance, they used 'big' instead of 'tall' and 'small' instead of 'narrow'. In the slow-learning groups there was a wider variation in the understanding of concepts than in the able groups. Several children, including some of the able ones, had an inadequate knowledge of number facts for the written calculations they were expected to do. Often, the teacher was unaware of the extent of each child's recallable number knowledge. This would require attention during the second input: teachers needed help both in analysing essential number knowledge and in assessing what the children knew.

The able children had little difficulty in performing and explaining written calculations and in developing and using new methods. Very few slow-learning children could carry out written calculations successfully. Those who could do so were usually able to explain their methods. The success rate in the slow third and fourth year groups of the middle schools with two-digit calculations was: addition 70 per cent, subtraction 30 per cent, multiplication 15 per cent and division 0 per cent. This suggested that slow-learning children had little knowledge or understanding of the four operations after a diet of little else for four to five hours a week over a period of seven or eight years.

(There must be something more productive for them to learn in mathematics.) They had enjoyed probability activities during these sessions; their success gave them an incentive to learn some number facts. I found that when concentrating on a few related number facts at a time the slow children made efforts to learn and were no longer daunted by the vast number of facts they were expected to memorize. I decided that it was important to persuade teachers to help slow children to *use* mathematics so that they could see a purpose in what they had to memorize.

On the other hand most of the able children were not given sufficient stimulus by the textbook examples which most of them were set. Although they sometimes worked with a friend they were rarely given the opportunity for discussion, either with their teacher or their peers.

3. All of the slow-learning children disliked and feared mathematics. They not only lacked confidence in their ability to learn the subject but seemed to have no expectation of understanding it. Most of them improved in their attitude to learning mathematics after the third session, but some of the disturbed children continued to be unpredictable from time to time. They were prone to be more affected by external factors such as the venue or an upset at school. When their attention began to wander (often when the session took place at the end of the afternoon) a change of activity usually alleviated this problem.

    Some of the able children also said that mathematics was dull. They commented 'It's the same old things all the time, day after day, year after year.' They needed to experience mathematics as a subject not only concerned with the acquisition of calculating skills, or even with problems associated with calculations, but with pattern in number as well as in shape and with attractive presentation.

4. All the teachers asked the children from the groups to talk about what they had done after every session. Some of the children were asked to teach a game or activity to other children; sometimes the teacher used the activity herself, as I had hoped.

These visits certainly enabled me to keep abreast of the changes which were taking place in the teaching of mathematics in the project schools. This information was usually picked up in the staffroom, when a teacher asked for advice, often about how to continue a topic he or she had started.

Outcomes of the second interviews with the heads and the teachers

These interviews took place at the end of the summer term, 1977. My intention was to discover the views of those heads and teachers interviewed previously about the impact of the project so far, and what they thought would be the most useful components of the second input.

The remarks made by the head of a first school (with the on-site pattern of working sessions) were typical of that group of schools:

> The working sessions went rather too quickly without time to consolidate one topic before moving to another. More time for discussion and the fitting of new ideas into place in our own scheme may have been helpful.

The major concerns of the teachers from first schools with the centre-based pattern of working sessions were progression and place value. One of the heads made another suggestion which I was able to follow: the teachers should be asked to choose one aspect of measurement to plan in detail and to try out with their classes.

At the middle schools, also, there was anxiety about the concept of place value and about planning for progression. Some teachers asked for more activities in fractions and decimals. All emphasized the need for help with progression throughout the age range. It was interesting to notice that this time, a year after the first input, there was no mention of the confusion caused by the earlier working sessions.

The second interviews confirmed the priorities I had selected.

The second input

The working sessions

The teachers now appreciated that the organization of the working sessions provided a model for their own classrooms. The atmosphere was positive: teachers received encouragement for their efforts. When they made a mistake they were questioned about their methods, but they were never criticized. (Any mistake might well have been caused by my teaching!)

A major purpose of the investigations given, and the questioning which followed, was to promote learning without telling the answer. Moreover, working in groups facilitated interchange of ideas and comparison of methods.

Three working sessions and two support visits were planned to take place in the autumn term of 1977 for each school.

For this input the working sessions for the centre-based schools were organized in phase groups, not area groups. The heads were also invited and five out of the eight heads took advantage of this. The total time allowed for the three working sessions was under five hours; the duration of each session was little more than half that of the first input, to enable teachers to be with their classes during the first hour of the afternoon. I adhered to the teachers' preference for topics: place value activities and the preparation of a sequence of experiences in volume and capacity, the aspect of measurement chosen by all the teachers.

The attitude of the teachers during the second input of working sessions was receptive and relaxed. They realized that they had problems in common in their schools when they were implementing changes, more especially when they were trying to help their colleagues to change. They voiced their satisfaction at meeting teachers from the same phase, but working in the other area. There was no longer an atmosphere of criticism or any reluctance to contribute, but a sincere desire to compare problems and successes. I wondered what had caused this change: a gradual increase in confidence in their ability to teach mathematics by means of planned activities? The knowledge that I was always available to help during the support visits but never to pressurize? The shortened sessions? Whatever the causes, the sessions were well attended and a good pace was established and maintained. The teachers were frank in their criticisms and discussions of conditions in their schools which inhibited them in their experiments. Most of the teachers brought some of the new work they had done with their children, attractively recorded by the children themselves. Several of them had enlisted the help of their colleagues in this exercise; one middle school team brought a complete sequence of work from the four-year course.

Because none of the teachers had had practical experience themselves of volume and capacity, the topic they chose to plan for progression, the initial practical activities which they chose to carry out themselves left less time for planning than I had expected. But the teachers judged their co-operative effort to be well worthwhile, although their planning was incomplete. They said that they now realized the value of involving all the teachers in their schools in preparation of this kind as a basis for making a school scheme in mathematics.

The positive reactions of the teachers to the second input of working sessions suggested that it was important to have two separate inputs. The change in attitude of the teachers between the two inputs was striking: the pace had increased and there were fewer critical comments.

At the working sessions of the schools with the on-site pattern of in-service education the atmosphere was more relaxed than before, and the teachers were more co-operative – but not to the same extent as were the teachers at the centre-based schools. Children's work was contributed from every class; much of this showed careful progression. When the working sessions were different in any way, the differences will be seen in the following reports on individual schools.

Support visits: progress and the teachers' comments

Centre-based first schools: Frame, Fleet, Flanders and Fowler

By this time the heads and the co-ordinators were fully apprised of the purpose of support visits and prepared a programme in advance of each. The programme included every teacher (except at Fowler). Two of the schools continued to make maximum use of the visits. The teachers' comments made during the support visits give an indication of their views of the changes which were taking place in individual schools; these are included in the following sections.

Frame

There was still no co-ordinator. Once more, at both support visits the head asked me to work with all the teachers during the lunch break.

One session was on place value, the other on suggestions for 'talking mathematics' lessons with children.

The teachers who had been trained overseas were finding group activities difficult to implement successfully. They found it hard not to ask questions which told the children the answers to any investigation they had provided, but they made efforts to introduce class 'talking mathematics' sessions. The head commented on the effect of the project so far:

> The project has been good in making the staff think and discuss. This did not happen before the project.

One key teacher said:

> The project has helped me to structure my lessons with the language and the games.

Another key teacher commented:

> I am enjoying my young reception class . . . But the project has confused me. It is difficult to introduce the ideas in a rather traditional school. But a positive achievement has been the games for children. I would not have thought of games without the project.

(This teacher had complained at the beginning of the project that games took too long to use.)

Fleet

The second co-ordinator, a former junior school teacher, had a good mathematical background and a liking for the subject. Like the first co-ordinator she had a preference for class teaching in mathematics but the head was 'working on her'. Because numbers in the classes were increasing, the head had not found it possible to free the new co-ordinator to visit the classrooms of her colleagues. But she realized the importance of such visits and promised to organize some non-teaching time for the co-ordinator in the future.

All the teachers had been trying out the new number scheme they had prepared together. The head was usually to be found helping the teachers or the children. At the support visits, as before, the head and the teachers decided on a specific topic in advance. No time was wasted and every teacher was included in the programme. The head commented on the value of the project:

> The project has created an awareness of mathematics; everyone has become involved. There is more mathematical comment now.

The second co-ordinator added:

> Before the project we did not see each other. Now we have conversations [about mathematics] informally.

Flanders

The head attended all three working sessions of the second input, but this did not give her enough confidence to offer help to individual teachers. In consequence, the co-ordinator did not receive the active support which might have given her the impetus she required to help her colleagues, so no progress was made. Neither did she begin to work on a scheme for mathematics. The deputy had definite ideas about the limitations which should be put on written calculations for the children. Yet paradoxically she gave her third year children 'six sums to satisfy the parents'. Another teacher who was near retirement was a 'passive resister'.

The head continued to vacillate about arranging meetings for me to work with all the teachers, yet she volunteered that 'the teachers do not expect enough from the children'. The co-ordinator commented that 'the project has made me think. I'm more maths orientated. I no longer say: "This is how you do it!"' Yet she continued to write most of the work for the children on the blackboard.

Fowler

The head made it clear to me that she thought the project had continued long enough. The first co-ordinator and one key teacher had left and the other had applied for transfer. The new co-ordinator had read widely to prepare a realistic scheme: not too long to discourage teachers from reading it, but sufficiently helpful to encourage them to make a start on necessary changes. She had also reviewed the equipment and ordered the material required to implement the new scheme. The head had given her some non-teaching time for preparing it and for assessing which teachers were in need of help.

Many teachers at this school needed guidance. They had been totally reliant on the workbooks of the former scheme which the head had decided to phase out. The co-ordinator tactfully told her colleagues that there was no longer any money available for the workbooks. She gave them all a copy of the new scheme but she did not have opportunities to help them to use this to best advantage.

At the support visits (limited as before to mornings only) I worked with the co-ordinator and with a few young teachers who had difficulty in controlling their classes. The head continued to be wary about my contact with the other teachers.

The co-ordinator (who had missed the first input) showed a marked change of attitude to teaching mathematics at the end of the second input. At her first interview she had volunteered that the first time she had enjoyed mathematics was when reading a book on the fascination of numbers. She said:

I was badly taught at the secondary stage and learned without understanding. I therefore find it hard to accept that understanding is important for children. Why not just tell and practise? It was successful with me.

Later on, she added:

Children are not challenged enough and do not have a secure knowledge of number facts. They cannot answer simple everyday problems.

After that interview the co-ordinator tried an alternative method of teaching essential number facts. She prepared attractive number games which she adapted from various sources. These were put to excellent use with her third year class, even with the slowest group. The games were well followed up; as a result all the children learned the essential number facts. This experiment was a great success. I realized how valuable this co-ordinator's work would be as a source of in-service education for other teachers in the school.

First schools: on-site pattern of in-service education
Foster and Finlay

The two schools, Foster and Finlay, made an interesting comparison. In some respects they were in marked contrast, in others similar.

Foster had a new head; Finlay had a long-established head. Both schools had recently moved to new, open-plan buildings (supplemented by huts at Foster). There was a limited amount of co-operative teaching at both schools. Apart from the departure of the first co-ordinator at Finlay both schools had stability of staffing during the first two years of the project. Unfortunately, however, one recently appointed member of staff at Finlay had an unsettling influence on the others (and three teachers left at the end of these two years).

As far as mathematical background was concerned there were further differences. The head of Foster had an adequate mathematical background herself; because she had a co-ordinator who preferred teaching older children and who was unable to help her colleagues, the head had undertaken to implement the project in this school herself. She achieved this in three ways: by helping all the teachers in their own classrooms; by working with groups of children herself in their classes; and by enabling all the teachers to assess two children at a time while they were working on practical investigations. Furthermore, the head had worked out a scheme for mathematics of which every teacher had a complete copy. In addition, she had held a practical session in mathematics for the parents.

By contrast, the head of Finlay had an inadequate mathematical background. Her professional course at college had done nothing to remedy this. Although this head gave every encouragement to the mathematics co-ordinator and to me, she was not able to offer help to teachers in their classrooms. Moreover, although the second co-ordinator was a knowledgeable and imaginative teacher himself, he too was unable to help his colleagues, partly because of the overall organization, which meant that he had to be with his class all day. The head of Finlay did not, at that time, arrange a meeting for the parents, although some of the teachers defended their traditional methods of teaching mathematics by insisting that 'the parents want to see sums – and so do I'.

With help from the head, the teachers at Foster adapted more easily to team teaching (indeed the two teachers from a junior school who had taught for the first year in classrooms volunteered, at the end of that year, to share the vertically-grouped first and second years in an open 'bay') and were willing to experiment. They felt supported by the head in any changes they made. She set them an example by taking new activities in their classrooms with groups of children. Moreover, the assessments they administered regularly to children gave them training in questioning rather than instructing. On the other hand, although Finlay had better facilities for team-teaching, some of the teachers were more set in their ways and more anxious about experimenting. Although the co-ordinator set a good example by his own teaching, the teachers were not able to see him at work. Neither were they given active encouragement to make changes in their teaching.

In view of the differences between the two schools it was not surprising that working sessions and support visits developed differently at the two. Both schools had resisters: Foster had three such teachers at the outset; Finlay had two. Nevertheless, the atmosphere of the working sessions at both schools was more co-operative than during the first input. The most negative teacher at Foster now had a first and second year group of children by request; she therefore felt more insecure and asked for help at the support visits.

The head of Foster who was doing so much herself to promote an improvement in the teaching of mathematics always made maximum use of my time at the support visits. The sessions were longer than those during the first input to allow for further development of the topic chosen. The teachers provided the material and prepared the activities in advance; time was allowed for appraisal and discussion afterwards. One of the two teachers in a team situation would join me to observe the children, to listen and appraise my questions and the children's responses. The head said that she, too, had changed her attitude to mathematics and her assessment of its importance.

The teachers at Foster were beginning to realize the value of 'talking mathematics' sessions and the need to ensure that the children not only understood the language patterns but could use these themselves. Perhaps the regular practical assessments they used with pairs of children were convincing them. It was clear that these teachers were now helping children to develop more than one method of subtraction.

At Finlay the teachers were not yet convinced about the value of organizing discussion sessions on mathematics for groups of children. The children at this school seemed to be over-dependent on their teachers and 'clamoured for attention' as soon as the teacher settled with a group.

The head, who was anxious to get talking sessions under way on a regular basis, saw no difficulty in timetabling a welfare assistant to help with such sessions. However, they never materialized, because the teacher concerned was always busy with something else when the welfare assistant arrived, and she went away discouraged. Was this symptomatic of the attitude of the teachers to the changes proposed? When I asked one of the resisters what she had been doing in mathematics during the two terms in which I had worked with groups of children her reply was: 'Oh, I've not thought about maths since you went'. On more than one occasion this teacher asked not to be included during the support visits. The work I did with her class on support days was never followed up, and sometimes the materials required were 'not available'. 'I cannot stand anything messy', this teacher remarked. Yet she applied herself well at the working sessions and did not oppose the project overtly in any other way.

All the teachers made an effort to bring to the third working session children's work from a sequence of activities. The two teachers from each year group had co-operated over this. The children had been encouraged to present the work in an attractive way; there had been a good deal of discussion among them. The 'resister' had brought a tape recording of the children's conversation.

However, remarks made to me, or to the head and repeated to me, show that some teachers were conscious of the changes which were beginning to take place. At Foster the head said:

> There is now more equipment in use and the teachers are talking to the children more during mathematics.

The deputy, a former 'resister', said:

> The project has made me aware of the importance of language in every part of the classroom . . . I have a good deal more confidence now . . . although we did not agree with everything [you] did.

Although the head of Finlay felt that the project had had much effect, these views were not supported by all the teachers. Two resisters said:

> Because I am unwillingly working in an open-plan situation I could not put many ideas into practice. Certainly there has been a change to some extent . . . I think we do enough talking maths now.

The other, a teacher of fourth years who relied substantially on workbooks, said:

> Because I like silence in the classroom I talk [maths] to the class between 9.00 and 9.30 every day.

(She admitted that the children did not have an opportunity to 'talk maths' themselves.)

The co-ordinator at this school remarked:

> The project has changed my teaching in many ways. It has expanded my repertoire and enabled me to adapt ideas, taught me the importance of language structure, helped me with content. We've spent time on making a record sheet.

This was the first occasion on which the co-ordinator had met the teachers officially to discuss mathematics.

Summary of the progress of the first schools

It seemed evident at this stage that all the second round of working sessions for the first schools were fully used by the teachers, who now appreciated their value as an opportunity for exchanging ideas with their colleagues, for learning more mathematics and for experiencing the planning of a progression of activities to help children to acquire concepts. But the overall time allowed for the working sessions was insufficient. The support visits, too, were well used, and on the whole well planned. At those schools in which the head gave active support in the classrooms a steady rate of change in the teaching of mathematics was evident, particularly in the planning of activities, in the attention given to questioning and in the amount of talking. When the head did not feel able to give active support, the effort demanded from the teachers was greater.

Support visits to centre-based middle schools

Centre-based middle schools: Melia, Meakins, Movehall and Makewell

The heads and the key teachers of these middle schools were determined to make the most of the support visits. Teachers who had not yet been involved in the project (including some appointed since the first input) were persuaded to request my help in the classroom. Furthermore, I was frequently asked to organize a practical session on a specific topic for all the teachers, or for a group, after school. Sometimes I was consulted by individual teachers about a mathematical concept they did not understand, or about planning the continuation of a sequence of activities. The accounts which follow indicate the extent of the changes which had taken place up to that time, as seen by the head and the teachers.

Melia

There was a high staff turnover in 1977. Of the sixteen teachers, seven, including the co-ordinator, left and were replaced. The new co-ordinator showed a lack of confidence in some respects. She wanted the children to enjoy mathematics but found this uphill work. She was afraid of seeming to interfere in colleagues' classrooms, but was eager to help them.

The fourth year children she taught appreciated what she did for them. 'She is a very good teacher of mathematics,' they told me spontaneously. She worked well with the other two key teachers.

In supporting some of the teachers in their classrooms, including those who were new to the school, the head came to realize how many of them were insecure in their teaching of mathematics. He therefore introduced a commercial system of workcards in the first-year classes as an experiment. The two teachers concerned were initially lacking in confidence; they worked through the cards during the summer holidays. Both began to gain confidence as they used the cards. One admitted: 'I am confident when teaching maths for the first time in my life.' This development had two consequences. The head began seriously to consider whether he should gradually extend the workcard system throughout the school. Meanwhile, the remainder of the teachers, hearing the enthusiastic comments made by their colleagues, expressed a desire to use the cards themselves. But the key team were reluctant to use them because they were developing their own scheme. The head asked the key team for their views on the card system and also asked me for my opinion. I recognized the value which an increase of confidence had for teachers with a negative attitude to mathematics. On the other hand I knew the disadvantages of an individual scheme: teachers often become administrators (marking answers, answering questions and recording progress) and the children suffer from a lack of adequate contact with the teacher. Moreover, to make the scheme effective for all the children the teachers need to be selective in the use of the cards. But in the trial year, they naturally tended to use every single card.

This discussion unfortunately left the head with the impression that I was totally opposed to the workcards, and tension grew up between the head and the key team, who were reluctant to use the cards. The head became determined that *all* the teachers should use the workcards. The key team queried: 'What was the use of sending us to working sessions which stimulated us to prepare practical activities to cover the essential concepts we think our children should learn, and then direct us to introduce another scheme altogether?' I had to explain to the key team a way in which the cards could form the basis for their course without being used in their entirety. (The head had not suggested that they should rely only on the cards.) If they showed goodwill by using the cards once or twice a week, these could provide material for the practice sessions which children undoubtedly needed. The teachers could then develop the concepts as they wished. Once the key team accepted the situation, the tension abated and I was able to convince the head that I was not opposed to any textbook or workcard system in principle, but only anxious that these should be used to the best advantage by the teachers for the children.

The head continued to give help in the classrooms to those teachers who asked him for it; the co-ordinator gave assistance to other colleagues. I worked with any teacher suggested by the head or the co-ordinator. One of the fourth year teachers was loath to teach mathematics to her least able set. I spent a good deal of time with her, starting activities with the children and leaving the teacher to continue them. But the teacher told the head that I confused her; the activities were never followed up. She was also unable to accept help from either the head or the co-ordinator.

Six teachers at this school attended a mathematics course organized by the LEA advisers. Their interest was partly stimulated by the key team, but the promise of full support by the head undoubtedly contributed to it. Moreover, the head continued to ask for working sessions after school for all the teachers on support days. In addition, the head informed the parents about the objectives of the project and the extent of the co-operation the teachers were giving.

The comments written by the head and the key team about the project underline their views. The head wrote:

> The sheer enthusiasm of [the researcher] and her counselling of the less confident staff members has probably had even more good effect than the expertise so widely disseminated. All of this has made it easier for me to provide help where needed. Scarcely anyone, even new appointments, has not been influenced.

The second co-ordinator wrote:

> My own mathematical concepts have been considerably widened by the workshops and doubtless will go on increasing (hopefully?).

The newest key teacher wrote:

> I dropped out of maths at the secondary stage. I'm 100 per cent with the project. Your approach suits me [very new to teaching] very well.

The other key teacher said:

> The project has re-structured my teaching.

Meakins

The head had always had a good attitude to mathematics, yet, once she had appointed a co-ordinator, she refrained from intervention and from active support of the teachers. On one occasion she said to me: 'I wonder whether I should have taken a more definite lead? Do I expect the teachers to be more imaginative than they are?' Although the question seemed rhetorical, I replied that the head's suggestion might well be true. I wondered whether her secondary teaching experience had caused her to adopt this stance of non-intervention in the teachers' work within their classrooms. She was frequently in the staffroom, where she always showed great interest in the teaching. Perhaps the head's attitude accounted, in part, for the relatively limited amount of change which occurred in the teaching of mathematics. There were several teachers in their first posts who would have profited from classroom help.

Moreover, the co-ordinator continued to be reluctant to assist his colleagues. This was understandable as far as the few much older colleagues were concerned, but not for teachers in their first posts. Once again, the co-ordinator appeared to vacillate about his duties. However, he had prepared a scheme for mathematics, which, after discussing it with the head, he presented to the teachers 'in a practical way'.

The two most experienced teachers in the school did not ask for help at the support visits after my first session with them. There were, of course, some young teachers who regularly asked for help with specific topics. They always took these topics further with the children between the support visits. Of the effect of the project the co-ordinator wrote:

> The staff are thinking more about mathematics and are aware of what they are doing. I think more about the way I plan my work. I know, too, that children do not necessarily learn from one good lesson. They may progress over a period of time. Therefore I must plan carefully.

The influential teacher who always adapted activities from the working sessions and passed these on to her colleagues wrote:

> The project has helped me very much indeed – by causing me to analyse my own teaching methods, giving me new ideas. The games sessions have been so helpful.

The second key teacher, whose change was more gradual, wrote:

The project helped me by giving me various ways of introducing different topics in more interesting ways.

Movehall

Three teachers out of eight, including the co-ordinator, had left the school before the second input. Perhaps because of the informal teaching style which was used in this school, the head and the teachers expressed interest at the support visits in the applications of mathematics to other aspects of the curriculum. But at that time, despite the encouragement I gave, there was no development of the starting points discussed. The head was now teaching mathematics himself to a third year set. Most of the material he used was taken from textbooks.

All the young teachers had many discussion points they wanted to raise at the support visits. It was not always possible to give them the assistance they needed, since mathematics appeared on the timetable only two or three times a week. Of the effect of the project the head wrote:

> You have helped in the personal development of the two teachers who went to your working sessions. One has now been appointed co-ordinator for maths in the lower team. The other teacher will shortly take up a similar position in the upper team.
>
> Your support visits have given an increased awareness of the need for structured but practical teaching of mathematics. It has been particularly valuable in developing oral work as opposed to recording for the sake of recording. The working sessions in 1977 appeared to be far more valuable to staff compared with the previous year, I noticed far more spin-off into the school.

I agreed with these comments. The two key teachers had been too preoccupied with trying out, in their own classrooms, activities derived from the working sessions of the first input to be able to pass these on to their colleagues. By the second input, both had gained confidence and were able to discuss the material from each session on their return. Moreover, they had persuaded all their colleagues to undertake a sequence of number activities with their children. So, by the second input, these two key teachers had taken up their roles, and were ready to act as joint co-ordinators. The two key teachers wrote:

The project has helped me and made me more confident to use books.

The project helped me an awful lot personally. I now do much more talking [with the children].

Makewell

The impetus of the first input was maintained throughout the second input. By this time the head was teaching an able third year set himself. He often discussed what he was doing in the staffroom, which stimulated the teachers to further efforts. The co-ordinator, who had missed the first input, attended the working sessions of the second input. She was very persuasive with her colleagues; she was released from some of her teaching to work regularly with new teachers who required help. The head also supported teachers in this way.

The head and the co-ordinator had clear views of where I could give most assistance. I was asked to work with the most and least experienced teachers at the support visits. Two experienced teachers who made no attempt to change their teaching of mathematics were diverted to other aspects of the curriculum. There were also some teachers who lacked confidence but were extremely anxious to learn more mathematics; they received full support from the head and the co-ordinator. The head wrote:

The majority of the staff have made efforts to change this year.

The co-ordinator wrote of her informal methods:

I feel I am now more relaxed. No longer feeling it necessary to race through a topic. But let the bright children work quickly then pursue 'advanced' work, letting the less able have more time to consolidate a concept.

Effect of the course on the staff

The key teachers [all new] are competent generally in the class-room and very keen to teach in the manner best suited to their pupils. They are both able and willing to change. When in doubt they will seek help.

In this school all discussion seems to take place in our somewhat confined staffroom. This lack of space helps mathematical development in that once [x] and [y] begin to chat with me, or each other, others are inevitably drawn into the conversation. If a game is demonstrated others will watch and comment. We achieve far more by informal discussion, arising from the course, or general problems, encountered in mathematics teaching, than by any formal meeting. We have year team meetings – smaller units than a staff meeting – where staff are encouraged to tell any other members of their team of any new successful approaches to the teaching of any subject. Mathematics is of course included.

The co-ordinator's comments have been quoted in full because this school was one of the most successful in improving the teaching of mathematics, despite a high staff turnover. Yet the in-service methods used were mainly informal and unstructured. Because a new commercial mathematics scheme had been introduced there was no written scheme for the school at that time. However, the co-ordinator was beginning to have doubts about the suitability of some of the material. She began, for the first time, to think that a written scheme might be necessary.

Middle schools: on-site pattern of in-service education. Missingham and Measures

Once more, the two schools receiving school-based in-service education make an interesting comparison. The heads of both schools had good mathematical backgrounds and enjoyed the subject.

The head of Missingham had been appointed to this, her first headship, after the school had been without a head for a term. Although she was supportive of the project she was not able to offer active support in the classrooms. The deputy head, appointed in April, 1977, had been responsible for mathematics in her former school, but she was reluctant to add this responsibility to her existing load. When persuaded during the second input, against her will, to become the mathematics co-ordinator, she chose to organize the equipment first and to leave the making of a scheme until later on, when she had had time to study the teaching of mathematics in the school.

The head of Measures had been at this school for ten years and was nearing retirement. His experiences with the third year set he taught caused him to have frequent discussions in the staffroom. This interchange with the teachers was a useful source of in-service education for them.

Although no time could be allowed for the co-ordinator to give classroom support to individual teachers, she was determined to effect improvement in the teaching of mathematics in the school. So, whereas the teachers at Missingham had hardly any day-to-day support when implementing activities suggested by the project, the teachers at Measures had some support (although not in their classrooms) from both the head and the co-ordinator. Furthermore, although Missingham had no mathematics scheme, the co-ordinator of Measures had already prepared a scheme which, after discussion with the head, she had presented to all the teachers.

The development of the topics at the working sessions differed at the two schools. At Missingham, at the suggestion of the new co-ordinator (who thought that the teaching of mathematics at the school was too formal), the sessions began with a discussion of the types of organization which would facilitate the introduction of activities and discussion. Attention was later focused on place value and the preparation of a sequence of activities on volume and capacity. The working sessions at Measures began in a different way. On the first day of the school year a written mathematics test had been given to the new entrants from the first school. The head (and I) had been doubtful about the wisdom of giving a test at that time and also about the knowledge which would be gained by the teachers. In the event, the test had proved unsuccessful in its purpose – to allocate the children to sets according to their achievement and ability in mathematics. The teachers were crestfallen about the failure of the test and asked for an alternative at the first working session. I suggested that group activities in probability might provide the teachers with an insight into the extent of individual children's understanding of number concepts and the extent of their number knowledge. Such experiments would provide a new start for the children rather than a depressing reminder of facts they had forgotten. We then carried out experiments in probability which also served to revise some of the activities included in the first working sessions. Time was spent on planning the possible uses, with children, of a collection of car numbers made by the teachers. The wide range of the activities they suggested perhaps indicated how much the teachers had learned from the working sessions. In view of the different degrees of support available within the two schools, it was not surprising that improvement in the teaching of mathematics proceeded at different rates.

Missingham had 13 teachers and Measures had 22 in 1976. Both schools initially had the same number of first appointments in need of help. Six teachers at each school left and were replaced between the beginning of the project and the end of the second input (five terms). There were teachers who resisted change at both schools: three at Missingham (two experienced) and one at Measures. But the number of resisters at Missingham was nearly a quarter of the total number of teachers. (One of the resisters did change – as a result of the new workcard system.)

Changes in the teaching at Measures proceeded steadily, with one exception, although initially I had the impression that the teaching methods at this school were more traditional. The head's objective (to provide the children with a quiet working atmosphere) appeared to facilitate the change to group activities.

At Missingham there were more teachers with disciplinary problems. They had to expend great efforts to change successfully to using mathematical activities with children organized in groups. (The head provided support for teachers experimenting in this way.) But the teachers at Measures had an additional advantage. Although they did not have opportunities to see the head or the co-ordinator at work, they were well aware that both were making changes in their own teaching styles. However, the co-ordinator had doubts about whether some of the teachers at the lower school (in buildings across the road) continued their efforts in my absence. This caused me to redouble my attempts to initiate extended activities to be completed before my next visit.

The teachers' comments made during the support visits, or written as part of the assessments of the project so far, also illustrate the difference in the rate of change. At Missingham the head said:

> The project has made the staff more aware of the importance of practical investigations and of language. But the work (at this school) is not integrated – staff lack a scheme so that there is no progression.

One teacher in his first post, a 'resister', told me:

I believe in attending to the four rules first.

The methods were demonstrated on the blackboard, after which the children worked from textbooks. He always had difficulty in controlling the class but made no changes because the children never reached proficiency in the operations. Neither the co-ordinator nor I was able to help him. The head of Measures had said of the project:

> The teachers have asked if we could have something similar in English. Mathematics is now on the mat — not swept under the carpet. The project has had an effect on other subjects, too. I am convinced that on-site working sessions have been more effective than off-site sessions. The support visits have been even more important. Some teachers were timid at first but the project has changed the temperature — the emphasis has changed. [Your] presence has helped good teachers, too.

The head later wrote:

> The project has stimulated much thought and discussion. Teachers who were previously diffident about discussing mathematics because they felt it was their 'weak' subject, found common ground on which to base discussion. This has also resulted in teachers having a wider knowledge of what others are attempting in their own groups.

The co-ordinator said:

> The project has given more ideas in a practical way. The first working sessions were fragmented. We now know more.

The comments of some of the teachers emphasized their problems as well as their successes:

> Debbie showed delight when she succeeded in achieving something. I would not have noticed this if she had been one of a class . . . I have tried to go very slowly on the basic fundamentals so that teachers in other years can build on firm foundations and resist the temptation to 'race ahead'.

Other experienced teachers said:

> I've found I cannot always have the children quiet when I introduce activities. I have to change.

> Am struggling to relax the 'recording' aspect and concentrate more on the 'reasoning', via games, etc.

Summary of the progress of the middle schools

To summarize, four of the six middle schools were making steady progress in the changes the majority of the teachers were trying to make in their teaching of mathematics. At three of these schools the head set an example by his own active support of the project, by teaching a class himself or by working with teachers in their classrooms. In the fourth school it was perhaps the enthusiasm of the head which gave the teachers the encouragement they required; to his own mathematics set he gave most of the work from a textbook at that time. At the other two middle schools neither head took an active part in the project; nor did either school have a co-ordinator who was helping colleagues in their classrooms.

The changes the teachers were implementing took various forms: using mathematical activities when a new concept was introduced; providing opportunities for the children to discuss what they were doing; giving the children less work from textbooks; talking more about their difficulties to their colleagues.

The reactions of the high school teachers to the project

It had been my original intention to involve the mathematics teachers of the first year pupils at the two project high schools as far as possible in the work of the project. My aims were:

1. to establish closer contact between the teachers of the contributory middle schools and the mathematics teachers at the high schools;

2. to try to ensure a measure of continuity in the teaching of mathematics at the interface between the middle schools and the high schools;

3. to try, with this in mind, to help high school teachers to appreciate that there were other and more demanding ways of teaching mathematics than class teaching.

The heads of the two high schools were interested and co-operated from the outset. From time to time they organized meetings between the entire mathematics departments of these schools and myself. Usually the discussions centred on ways of determining standards reached in mathematics by the new intake which did not involve setting a test during their first week in school.

The head of the mathematics department at one school accompanied the mathematics teachers of the first year pupils to all the working sessions. Three teachers of first year pupils at the other school attended the working sessions arranged for the other area. All these teachers welcomed the opportunity of working with their colleagues from the middle schools. All took an active part in the practical activities and discussions which followed. Although the activities and language patterns were focused on the middle school age range, activities and investigations appropriate for pupils in the first year of the high school were also provided. However, on the whole, the high school teachers chose to work with their colleagues from the fourth year of the middle school. Perhaps this influenced the high school teachers' assessments of the working sessions. A teacher in her first post wrote:

> Course has proved interesting and useful to *me* but the work involved can rarely be used in high school due to pressure of exams and time. I would definitely make use of it teaching remedial groups at the lower end of the school (which I am not doing at present).

The teacher responsible for first year mathematics at the other high school wrote:

> It was interesting to see the general first/middle school approach to mathematics.

Despite these lukewarm assessments, both these teachers used some of the activities with groups of children at the support visits, but they did not continue this work on the grounds that it would take so long to see any lasting effect. Unfortunately, neither was able to visit the contributory middle schools to see the changes which were taking place in some of the classrooms. I also tried to encourage the high school teachers to provide investigations when introducing new topics, rather than giving class lessons which, however well developed, usually ended in instruction. Again, although one or two teachers who had not attended the working sessions did experiment in this way, the first year teachers turned the suggestion down, once more on grounds of pressure of time.

In high schools, as in first and middle schools, another problem arose. All except one of the teachers who had taken part in the working sessions were transferred to teach in other parts of the school and no longer taught first year pupils. In view of this lack of continuity and the continuing need of teachers in first and middle schools for classroom support, I decided to curtail future visits to the two high schools. I arranged to visit once a term to interview those children with whom I had worked for two terms and who had been transferred from middle to high schools. At the end of the summer term, 1977, I sent to the heads of mathematics departments at the two high schools the names of the able and of the slow-learning children in the groups with which I had worked in all the project middle schools. I indicated those children who had made outstanding contributions in the mathematical investigations I had provided. I also mentioned those children who were particularly apprehensive about the transfer.

A change of tactics

I had originally planned to make purely observation visits to all the project schools in order to assess the extent of the changes the teachers were making and where additional help was needed. When I discussed this proposal with some of the co-ordinators, they discouraged me. The co-ordinator of Meakins expressed his doubts as follows:

> You would make the teachers anxious; they would put on something special and not a normal lesson. You would learn more about what the teachers were really doing if you worked with them in your usual way. Anyway, after all the support you have been

giving you will find it impossible, now, not to take part. Teachers are resentful when they know they are being monitored.

These comments seemed reasonable to me, particularly since I, too, felt uncertain about the wisdom of pure observation at this stage. I realized how much more help many teachers still required to sustain the changes they were making. Moreover, in some schools there had already been a substantial staff turnover (often of key teachers). I therefore planned that my future visits to project schools should be support/observation visits and that I would continue to give support to teachers where this was needed.

CHAPTER 7

The contribution of the advisers

The assessment of the project: observations and support visits by the advisers

Introduction

In Chapter 1 I described the outcomes of the initial mathematics conference for all advisers in the borough organized in co-operation with the mathematics advisers. I planned to have two kinds of help from the advisers: one was to observe the changes made by individual teachers as the project progressed, the other was to support the teachers in their classrooms. All of the advisers and lecturers who volunteered their help had had experience of mathematics workshops, most of them over an extensive period, but only the mathematics advisory teacher had had first-hand experience of helping teachers in their classrooms to implement the aims and processes of which the workshops were an example. I had therefore decided that since the advisers were to make an assessment of the outcomes of the project, it was important to involve them in classroom support so that they would understand from personal experience what this entailed.

The nature of the classroom support which advisers would feel confident to give was discussed with the mathematics advisers. Since some of the volunteers did not have a strong mathematical background they could not be expected to provide the kind of support offered by specialists: to assist teachers to introduce new topics by means of practical activities. Other advisers could, however, support a teacher who wanted to work with one or more groups of children by taking responsibility for the remainder of the class. This form of organization would maintain order while the teacher observed the children as they tackled a particular activity, listened to their discussion and based her questions on their responses.

In the event five advisers and two mathematics lecturers volunteered to participate, by observing the teaching of mathematics in schools they did not know and by supporting the teachers in yet other schools. Although the mathematics advisory teacher was not allowed to observe, he agreed to give support in three middle schools in the project. The advisory team had agreed to give one support day to each project school, with the possibility of another support day during the second input. To minimize the calls made on their time, it was agreed that they should support (and observe) the key teams only. This would involve them in supporting three or at most four teachers in any one school. In fact the unexpected curtailment of the advisers' visits led to a change of emphasis during support days (and to a lack of continuity in the observations they made). With the exception of the advisory mathematics teacher the procedure in fact adopted by the advisers at support visits was to pay visits to all the classrooms and to follow this by a discussion about the value of the project with individual members of the key team. The written reports of these discussions were useful because they provided an independent confirmation of what the teachers had said to me at formal (and informal) interviews. In general, there was a remarkable degree of agreement between the key teams' assessments of the project at different stages to the advisers and to me. Sometimes the comments made to the advisers threw new light on a teacher's initial statements. For example, one adviser wrote of the teachers at Frame:

> The more experienced teachers feel that some of the subjects [concepts] are completely new, they never actually experienced them and they need a basic course for the underlying principles rather than the activities concerned with the new maths . . . As far as support was concerned this did not go far enough to help with organizing the creative ideas for work given.

This adviser arranged that a teacher in this school who was having difficulty with some specific apparatus should visit the other school in which the adviser gave support where she knew 'there was some very good work going on with the apparatus'.

The shift of emphasis during the shortened support visits meant that individual key teachers did not receive the classroom support they were expecting; neither did the advisers have the experience of helping them with groups of children for activities and questioning. The half day visits of the advisory mathematics teacher to three middle schools were an exception. (I spent correspondingly less time in these three schools.)

However, the adviser with responsibility for first schools was prepared to work with individual teachers in their classrooms. She wrote:

> Support is almost impossible done on the basis of a research project because one supports one's own investigations and observations and then one develops strategies appropriate to that school. Advisory teachers' sustained work with schools seems more fruitful. 'Support' for my support school will take a completely different form next term at [the teachers'] request and to my pleasure.
>
> 1. Work with co-ordinator.
> 2. Work with teachers.
> 3. Sessions with teachers.
> 4. Working with them in their classrooms.

Although the views she expressed were not altogether in sympathy with the research project, her plans for helping teachers were completely in harmony with the initial purpose of the support visits. It was unfortunate that LEA demands on this adviser's time increased to such an extent that she was unable to carry out her plans. To summarize: the support visits to schools made by the advisers were short and did not involve the advisers to the extent intended. Nevertheless, because the mathematics advisory teacher discussed his own support visits with his colleagues, at least they became aware, at second hand, of the nature and potential of such visits. As far as the teachers were concerned the advisers' support visits were of value because they made the teachers realize that the advisers were familiar with the aims of the project, were concerned in its implementation and were aware of the efforts the teachers were making and the problems they faced.

There was one respect in which the advisory team's knowledge was invaluable. The mathematics adviser organized meetings at termly intervals for the volunteer advisers, the lecturers and myself.

Their knowledge of individual teachers and heads, based on many visits (past and present), was of great benefit to me as a check on my own observations. I prepared my comments for each meeting but asked the advisers for their views before I revealed my own, to try to achieve a measure of objectivity.

Inevitably, the contraction of observation visits led to a lack of continuity in the observations. It was rarely possible for the same observer to follow the changes made by one particular teacher. Moreover, despite the initial expectation of staffing stability (because of the gradual reduction in the number of teachers employed) there was such a high staff turnover that making consecutive studies of individual teachers became impossible. The first observation visits were planned to take place before the teachers began to make changes, at the start of the first input. But the advisers found, as I had, that one visit was not enough to enable them to form an opinion about a teacher's style in teaching mathematics. For observations to be valid, visits needed to be made at intervals. The advisers had no time to make a second visit before the first input was under way and changes were already occurring.

In order to help the advisory team at their observation visits a schedule was prepared. For this purpose the mathematics adviser and I had a preliminary discussion about aspects which advisers without an extensive mathematical background would be able to observe. Together we drew up a schedule for discussion with her colleagues. When the final observation visits were made by the advisers in 1978–9, the advisory team had contracted still further. A revised schedule was prepared because of the limited time available for these visits. Moreover, as the project developed, new features became important. For example, partly as a result of the project and also because of the independent in-service education provided by the two mathematics advisers, one major activity in each school had been the preparation and trial of a mathematics scheme. Furthermore, the extent of the co-operation between the head and the co-ordinator had been found to be important.

It was unfortunate that because only three advisers took part in the final observations, they could spare no more than one morning at each school. This restricted the scope of the visit to discussions with the head and the co-ordinator and observation of one teacher. Although the records provided were not as extensive as I had hoped, they formed a basis for comparison with my own assessments made during many visits (at least 30, including interviews) to each of the project schools.

A comparison of the written assessments of the advisers with my own

The advisers' records rarely differed from mine to any great extent. Frequently these records reinforced my assessments of the changes taking place in a school. Usually because the adviser and I did not visit schools at the same time, their accounts interleaved with mine to provide evidence of the gradual changes which were taking place in individual schools.

Comparisons between my comments and those made by the advisers on one first school and one middle school follow. The first school, Fleet, was chosen because this was the only school for which the advisers' comments on one teacher differed from my own to any appreciable extent. I therefore had to consider additional factors when weighing up the evidence.

Fleet

The same adviser made all three observation visits. This was one of the first schools in which the head took full responsibility for assisting the teachers to make changes in the teaching of mathematics until she was able to choose and train a new co-ordinator.

The key teacher observed was a graduate in her third year of teaching. At her first visit in June, 1976, the adviser wrote:

> There are opportunities for discussion at all times. General conversational bustle all the time during my visit . . . The activities provided were too diffuse for me to tell accurately whether the teacher makes provision for all abilities.

In February, 1977, the adviser recorded:

> Although children in this class always seem very unsettled the thought which goes into the work is more imaginative. Nevertheless as yet it is not always fruitful because ideas are not sufficiently well-developed or executed.

A list of the interesting activities in progress was included. At about the same time the mathematics adviser visited to support the teachers. Her notes about this teacher seemed to agree with those of her colleague:

> . . . a lively, intelligent teacher who was keen to put the project activities into practice . . . Her class control was loose, resulting in a noise level too great for mathematical thinking and some children wasting time. With support she may overcome these difficulties and become a good teacher of mathematics.

My view of the current work of this teacher was more optimistic:

> This teacher was very encouraging to individuals. I think she knows the standards she hopes to reach. There was a great variety [of activities] in this large six-year-old class (nearly 40). With one exception the children were interested in their activities. Although they made many demands on this imaginative teacher they were becoming independent.

At my third support visit (October, 1976) I wrote:

> This teacher is highly organized and yet gives the appearance of allowing the children a good deal of freedom.

Because my appraisal was more positive than the assessments of the two advisers I was reassured to have the head's comments:

> A most promising young teacher. My only worry is whether she spends too much time on preparation.

I followed this young teacher's career with much interest. When she left (before the second input) she was given responsibility for mathematics at her second school (in another LEA) because of her work with the project. When she changed schools for a second time she was appointed teacher/consultant for primary mathematics, with responsibility for this subject in four primary schools. Considering the head's assessment of this teacher's professional qualities, and the subsequent development of her career, I was confirmed in my assessment of this teacher's work.

Melia

The head was convinced that working with all the teachers at a school would be effective in bringing about change; he therefore welcomed the support visits which would involve all the teachers. He offered to support his teachers in their classrooms himself. During the first input of the project he discovered that more teachers than he had expected lacked confidence when teaching mathematics. The extent of their negative attitude to mathematics was borne out by their own assessments: 67 per cent left both school and college with negative attitudes to this subject; only 33 per cent said that they were confident when teaching mathematics. There was another problem: a high staff turnover of 67 per cent which affected the key team as well as the rest of the teachers. The large majority of the teachers were willing to be helped and asked for help at the support visits. Enthusiasm was aroused and six teachers attended a mathematics course offered by the advisers.

Until the first co-ordinator left in April, 1977, the teaching of mathematics improved steadily, as the following accounts show.

At my first support visit in May, 1976, I recorded a definite change in the co-ordinator's own teaching, and in her work with colleagues, in comparison with my preliminary visits when I found her relying on worksheets. I recorded before the first input of the project:

It was surprising to find that the children showed so little enthusiasm for mathematics in view of the lively and encouraging manner of the co-ordinator.

In May, 1976, I wrote:

The co-ordinator is a lively person and an outstanding teacher. The children were working at a variety of activities, mainly concerned with number facts and properties.

Undoubtedly the co-operation of the head and the co-ordinator's enthusiasm are affecting the staff. The co-ordinator has already tried with her fourth year set all the appropriate activities used during the working sessions. She has helped and encouraged the key teachers and other members of staff to make the material they need (for activities and games).

One adviser made two visits for observation, in July, 1976, and January, 1977. After his first visit he wrote:

> I found the level of mathematics teaching to be a high one in comparison with other middle schools I normally visit. [The co-ordinator] is a particularly gifted teacher, all three groups were actively interested and working well according to their capacity.
>
> [The new key teacher] is to be congratulated on her good organization and use of space for a large and fairly ebullient first year class. There is good support from the head and maths resources within the school appear most adequate.

At a support visit in autumn, 1976, I commented:

> The co-ordinator emphasized that all the teachers would value help at the support visits. She is aware of the strengths of her colleagues and also where help is needed most. In the free periods allocated to her by the head she has been supporting those individual teachers in their classrooms who were willing to accept her help. She gained in confidence as a co-ordinator after attending the LEA advisers' conference for middle school co-ordinators.

At his visit to the school in January, 1977, the adviser wrote:

> Pupils discuss with their peers or with the teacher [co-ordinator] who is constantly observing. Able children are encouraged to use a variety of methods . . . including games which are self or group correcting. Individual help/group help from teacher.

The head was very appreciative of the other key teacher who had made rapid progress despite her original negative attitude to mathematics. Of her second year class I recorded:

> She has a variety of activities in progress and questions each group skilfully. She has already drawn up a scheme for her children based on all the work covered [at the working sessions]. She found this very satisfactory because she dislikes the textbooks in use and seldom gives practice from them. A promising young teacher who has overcome her initial fear of mathematics.

Summary of the comparisons

From these comparisons of the advisers' comments and mine of the progress of changes in the teaching of mathematics in two of the project schools, it can be seen that although the time given by the advisers was more limited than I originally expected, in general our assessments were in agreement. Because the advisers' contribution was restricted I relied more heavily on the assessment made by the heads. (This contribution is described in Chapter 8.)

A parallel experiment

The work of the mathematics advisory teacher

The mathematics advisory teacher was working in another 12 first and middle schools in the borough at the same time. His mode of in-service education was closely related to mine. Each had decided independently to work with teachers in their classrooms on a continuing basis. We had been unaware, in the early days of both experiments, that we were working on similar projects in different schools. We co-operated when workshops were taking place. From time to time we met to compare the progress of support visits to the schools.

Throughout the mathematics advisory teacher's service with the LEA his main emphasis was on the use of language in the teaching of mathematics. His order of operation was: Do . . . Talk . . . Record. He wrote in a paper which he and the mathematics adviser distributed to all middle and high schools:

> The children must . . . be able to explain the mathematical processes involved. It is, therefore, considered of the first importance to help children acquire the necessary language for such explanations before insisting on technical competence.

The aims of the mathematics advisory teacher were therefore in complete accord with mine.

One major concomitant of this advisory teacher's work with

teachers was his substantial contribution to centre-based in-service education. Several of the teachers in the 12 schools he visited on a regular basis (together with some teachers from my project schools) attended one or more of his sessional workshops. From his experience of in-service education he concluded:

> It seemed that both school-based programmes of INSET and teachers' centre-based programmes have been mutually beneficial to one another. This might suggest that both forms, at least, are necessary.

The mathematics advisory teacher was also responsible for the two conferences held by the LEA, one for middle school co-ordinators and one for those from first schools. He suggested that the aims of co-ordinators should include:

1. to involve all colleagues in drawing up a maths scheme for their schools;
2. to involve all colleagues in workshops, which they should initiate within their own schools, to suggest ideas for practical activities in maths;
3. to work alongside their colleagues . . . helping them to implement workshop ideas, the maths scheme, the maths model for teaching, and to achieve the language/technique expectations.

I attended almost all of the mathematics advisory teacher's working sessions at the teachers' centre and acted as his helper. In this way I became familiar with his way of working and was able to see some of the project teachers at work in a different setting. His concept of his function was similar to my concept of my own.

Initially he worked alongside the teacher in order to identify any pupil or teacher needs. Secondly, he set up learning situations in individual classrooms. At the final stage, 'the advisory teacher's role changes. He no longer instigates learning experiences, instead he helps the class teachers implement a plan which is essentially theirs . . . it is vital that the teachers are supported in this way. Innovation in the classroom inevitably involves setting up new patterns of organization and pupils have to be trained to operate them.'

Both of us identified the teachers' needs by working in their classrooms: the advisory teacher by working alongside the teacher, I by observation at that stage. Both planned learning experiences in consequence of our preliminary visits. At the initial support visits I, like the advisory teacher, took the initiative during the lessons planned jointly with the teacher who acted as a helper, taking responsibility for one or two groups. Gradually, as the teachers gained confidence, they took the initiative in implementing their own ideas. The mathematics advisory teacher and I believe, from this experience, that it is vital to support teachers in this way.

One difference in the organization of the two projects was the timing of the inputs. Both change-agents gave individual schools between 21 and 24 days of observation and support time. My programme was: first input, two terms, then two terms interval; second input, one term, followed by a diminishing number of support visits during the following five terms. The mathematics advisory teacher's programme was: first input, one term, then two years' interval (some teachers attended working sessions at the teachers' centre during this interval); second input, two terms. The mathematics advisory teacher, like me, found that some teachers said that they discontinued their experiments during the interval. However, both found that, after the second input and the subsequent support visits, changes in the teaching of mathematics had been adopted by many more teachers; the schools could then withstand a high staff turnover.

Implications of my findings and those of the mathematics advisory teacher

The two experiments have much in common. We both worked flexibly, adapting our methods as different problems arose. In both, school support proved to be a major influence in the changes the teachers made in their mathematics sessions. The advisory teacher first worked to persuade the teachers of the need to change. I also needed to persuade some teachers that change was necessary. Although I had the commitment of each school to take part in the project this did not mean that all the teachers were willing to co-operate; they, too, had to accept the need for change if this was to occur. In both experiments, once the teachers were committed to change, the initiative was gradually shifted from the change-agent to the teacher.

The time required to effect changes in teaching style was strikingly similar in the two experiments. It was perhaps significant that in most schools the 'tipping point' was not reached until at least three years after the project began, although the intensity of school support and the interval between the inputs were different for the two experiments.

But school support could not provide all the mathematical background which teachers required. The advisory teacher recorded:

As a result of working alongside teachers in their classrooms, it has been possible to identify some patterns of deficiency in the teaching of maths, for example, the need to make the development of the language patterns of maths a major objective for all in-service work at the primary stage of schooling.

My experience with teachers, both in their classrooms and at working sessions, had made me aware, too, of the need to provide structured experiences which would give rise to the language patterns of the operations. Classroom support was therefore supplemented, in each experiment, by working sessions with the teachers. The organization was geared to the changes the two change-agents were helping teachers to make in their classrooms. Informal groups of teachers worked at structured activities, and used and discussed the language patterns introduced.

The working sessions had another feature in common. The expression: 'That's wrong', was never used. Instead the teachers were asked to explain what they had done; in doing this they usually discovered where an error had occurred. At the end of the sessions the aims of the organization were made explicit to the teachers: that it mirrored both the classroom organization and the positive encouraging attitude we hoped that teachers would adopt.

Towards the end of his experiment the advisory teacher referred to the importance of both school-based and centre-based in-service education. It seemed to him, as to me, that to assist teachers to acquire the necessary background (for practical situations, language and additional knowledge) centre-based workshops were more economical than school-based workshops, because teachers from more schools could be involved at the same time. We also agreed about the importance of involving all the teachers at a school in the preparation and trial of a school scheme for mathematics.

The findings of the mathematics advisory teacher go a long way to suggest that my work could be replicated by others. Both experiments have so much in common that the results suggest that there might be a basis for generalization. Elliott (1980) wrote:

> Action research does not assume that its findings are generalizable. However, through the comparative study of cases it is possible to identify similar cases and therefore teaching problems shared by different teachers.

CHAPTER 8

The contribution made to the project by the heads, the co-ordinators and the key teachers

The contributions made by the heads and the key teams will be reviewed in this chapter, up to the end of the second input (December, 1977). Then further developments between 1978 and 1980 will be described. A final assessment of the outcomes will be made.

Introduction: the crucial part played by the heads

From the time of the first support visit I obtained the co-operation of the heads in identifying those teachers who would be willing to accept help in their classrooms. The heads were also able to advise me about special difficulties which might inhibit individual teachers, such as fear of losing control of the class, or a very scanty mathematical background. All the heads, except one, were consistently co-operative in this respect. As the support visits continued they succeeded in persuading additional teachers to ask me to work with them. In this way I gradually formed a picture of what the heads thought of the potential (and later on, the achievement) of individuals as teachers of mathematics.

At the end of the second input I asked all but one of the heads for their assessments of the extent of change in the teaching of mathematics. My familiarity with the heads' successive assessments was of particular value when it became evident that the advisers could not spend the time on observation which I had expected.

Responses to the written questionnaires made at the end of the second input

The questionnaires were distributed for completion by the end of the second input to all the project schools except Fowler. My intention was to obtain assessments of the effects of different aspects of the project (a) from the heads, (b) from key teachers at schools with centre-based working sessions, and (c) from all the teachers at schools with the on-site working sessions. The questionnaires, and summaries of the responses, follow.

Questionnaire for the heads

1. To what extent has the project had an effect on the teaching of mathematics in your school?   *No effect.   Some effect. Much effect.*
    Please delete as appropriate.

2. How many teachers have made changes in the teaching of mathematics in their classrooms during the past year? _____
    The total number of teachers concerned in the teaching of this subject _____

3. Which of the following aspects have helped your teachers?
    Please tick √ in the appropriate column:
    (i) not at all (ii) to some extent (iii) very much

	(i)	(ii)	(iii)
(a) Support from you as head			
(b) Help from the co-ordinator			
(c) Help from the key teachers			
(d) The working sessions			
(e) My support visits			

Please state anything else which has helped or hindered or anything else which I could have done.

Summary of responses from heads to questions 1 and 3(d) and (e)

(Expressed as a percentage of the total number of questionnaires returned. No head used the category 'No effect'.)

1.	*Overall assessment of the project*	Much effect	Some effect
	First schools	0	100
	Middle schools	50	50
2.	*Working sessions*		
	First schools	20	80
	Middle schools	83	17
3.	*Support visits*		
	First schools	20	80
	Middle schools	67	33

When the heads were asked to estimate the number of teachers who had made changes in their teaching of mathematics since the beginning of the project some difficulties arose at first school level. Only three heads of first schools were able to make this estimate; they assessed that, on average, 63 per cent of the teachers had made this change. The overall percentage for all six first schools would have been considerably lower, since the heads who did not reply gave little support within their schools. In the six middle schools the average percentage of teachers assessed by the heads as making changes was 61 per cent.

The heads' assessments of the effect of the project in their schools (question 1) also seemed to reflect the different degrees of support given within the schools. All five heads of first schools estimated that the project had had some effect, whereas three of the six heads of middle schools estimated that the project had had much effect; all three were actively involved themselves in giving support within their schools. Furthermore, the heads of the middle schools assessed the working sessions and support visits as being far more effective than did the heads of first schools. Few of the heads, at that time, considered that the co-ordinators and key teachers had given much help to their colleagues (questions 3(b) and 3(c)).

The criteria used by the heads

The teachers were judged by the heads to have made changes if:

1. they were providing more materials for the children to use (Frame, Fleet, Foster, Melia, Measures);
2. the children were talking more (Fleet, Foster, Melia, Movehall); a variety of approaches was used (Missingham);

3.   the children's attitude to mathematics had changed (Fleet); the teachers' attitude had changed (Frame – they talked of their problems and failures), (Foster and Fleet – they showed interest and enthusiasm), (Melia and Makewell – they asked the head for help); they worked less from textbooks (Missingham);

4.   they observed and listened to the children (Fleet); they helped the children to find an answer rather than instructing them (Missingham and Movehall).

The heads of Meakins, Makewell and Measures discussed with the mathematics co-ordinators the extent of the changes made by the teachers in their teaching of mathematics.

Did the nature of the criteria used by the head reflect the changes which the heads had made themselves?

Questionnaire for the teachers

1.   Please indicate if any of the following aspects have helped you in the teaching of mathematics in your classroom.

Tick   √ 1.   if not at all
          2.   if to some extent
          3.   if very much

In school		1	2	3
(a) Support and encouragement from the head	(a)			
(b) Help from the co-ordinator	(b)			
(c) Help from my support visits to your classroom	(c)			
(d) Help from any other source; please specify _____	(d)			
The working sessions				
(e) The content	(e)			
(f) Using materials and equipment yourself	(f)			
(g) Discussion with other teachers	(g)			
(h) Working with other teachers	(h)			
(i) The papers distributed	(i)			

2.   Have you during the past year made any changes in your teaching of mathematics? YES or NO. Delete as necessary.
If YES please indicate in what way:

(a) Giving children more opportunity to use materials and equipment. YES or NO.
(b) Giving children opportunity to talk mathematics. YES or NO.
(c) Not demonstrating how to do a piece of mathematics (for example a calculation) but helping children to work out a method for themselves. YES or NO.
(d) Working less from books. YES or NO.
(e) In the organization of your class, for example, letting children work in pairs or in groups. YES or NO.
(f) In any other way. Please specify.

Name   --

Teachers' assessments of the type of changes they had made (Expressed as a percentage of the total number of questionnaires returned.)

	Using more materials	Evoking more discussion	Organizing groups	Depending less on textbooks	Questioning rather than telling
First schools	59	69	54	38	23
Middle schools	89	78	73	70	51

Throughout these responses the assessments made by the teachers from middle schools were higher than those made by first school teachers. A possible reason for this discrepancy is that first and middle school teachers start from different baselines. First, there was initially very little material in use in middle schools for mathematics, whereas some material (for example, counting aids) was already being used by first schools. Secondly, middle school mathematics lessons were often silent, while first schools necessarily put more emphasis on oral work. Thirdly, since children at first schools normally did fewer written calculations, there were fewer opportunities for teachers to demonstrate the 'correct' method. Nor were mathematics textbooks or workcards so much in evidence. So in some ways first schools had less far to go except where, in these schools, teachers used more formal methods to teach mathematics than they used for other aspects of the curriculum.

Summary of responses from the teachers
(Expressed as a percentage of the total number of questionnaires returned.)

1. *The effect of different aspects of the working sessions*

	Content		Using materials for themselves		Papers distributed		Contact with other teachers	
First	some	81	some	85	some	81	some	65
schools	much	8	much	8	much	8	much	8
Middle	some	30	some	43	some	49	some	59
schools	much	65	much	54	much	41	much	32

2. *Relative amount of help received from support by the head, the co-ordinator, the researcher*

	Head		Co-ordinator		Researcher	
First	some	69	some	31	some	73
schools	much	19	much	4	much	19
Middle	some	54	some	27	some	41
schools	much	30	much	32	much	46

Note: Where the totals are less than 100 per cent the remainder consists of teachers who said 'No effect'.

Factors which facilitated change in the teaching of mathematics

As the project continued into 1978 certain factors which seemed to be necessary for promoting changes in the teaching of mathematics began to emerge. The presence or absence of these factors in each of the project schools is shown in Tables 8.1 and 8.2 and 8.3. The factors included do not have equal weight. The tables were constructed to help me to assess the overall contribution made by the

Table 8.1: The contributions of the heads.

	A	B	C	D	E	F	G	H	J	K	L	M	Total support 1978	1979
First schools														
area 1 — Frame		x					xR	x	x			x	4	5
area 1 — Fleet	x	x	x	x	x	x	x	x	x	x	x	x	12	12
area 1 — Foster	x	x	x	x	x	xR	xR	xR	xR	xx	x	x	9	12
area 1 — Flanders		x					xR	x	x		x	xR	4	6
area 2 — Fowler							x						1	
area 2 — Finlay		x	x		x		xR	x	x	x			6	7
Middle schools														
area 1 — Melia	x	x	x	x			x	x		x	x	x	9	9
area 1 — Meakins	x	x		xR			x	x		x	x	x	7	8
area 1 — Missingham	x	x				xR	xR	x		x	x	xR	5	8
area 2 — Movehall	x	x			x			x			x	x	6	6
area 2 — Makewell	x	x	x	x	x		x	x	x	x	x	x	11	11
area 2 — Measures	x	x			x		xR	x		x	x	x	7	8

Support for teachers
A Mathematics knowledge
B Co-operative
C Gives informal help
D Gives classroom help
E Teaches herself

Supports co-ordinator
F Trained co-ordinator
G Gives non-teaching time
H Provides staff meetings
J Provides parents' meeting

Supports project sessions
K Attended working sessions
L Attended children's sessions
M Active in scheme preparation

R After 1978

Table 8.2: First schools: Contributions made by co-ordinators and key teachers (and all teachers at on-site schools)

School		Co-ordinators										Total maximum 10	Number of key teachers	Key teachers (and all teachers at on-site schools)			Total maximum 3 per teacher
		A	B	C	D	E	F	G	H	J	K			L	M	N	
Frame		x	x	x	x	x	x	x	x	x		9	2	x x	x	x	4
area 1 { Fleet	(1)	x						x				2	3	x x x	x x	x x x	8
	(2)	x	x	x	x	x	x	x	x	x	x	10					
Foster	(1)	x							x			2	2	x x	x	x	4
	(2)		xR	xR	xR	xR	xR	xR	x	x		9					

Table 8.2: First schools: Contributions made by co-ordinators and key teachers (and all teachers at on-site schools)

		A	B	C	D	E	F	G	H	J	K	Number	L	M	N
Flanders	(1)	x						x				2	2	x	2
	(2)	xR	xR			xR		xR	xR	x		5	2	x	
area 2 **Fowler**	(1)											0			
	(2)	x	x	x	x	x		x	x	x	x	9	2	x	1
Finlay	(1)											0		x	
	(2)	x	x	x			xR	x	x	x		8(9)	2	x	2

(a) Co-ordinators
A Changed teaching

Standing in school
B Head confident in her
C Colleagues respect her
D Colleagues accept her advice

Meetings
E With staff
F With parents
G Contribution to teachers

Preparation of mathematics scheme
H Has mathematics knowledge needed
J Prepared by head and co-ordinator
K Prepared by all the staff

(b) Key teachers
Number
L Changed teaching
M Influenced colleagues
N Acquired more mathematics

R Recently after 1978

Table 8.3: Middle schools: Contributions made by co-ordinators and key teachers

School		A	B	C	D	E	F	G	H	J	K	Total maximum 10	Number of key teachers	L	M	N	Total maximum 3 per teacher
														Co-ordinators		Key teachers	
Melia	(1)	x	x	x	x	x		x	x			7	2	x	x	x	6
	(2)	x				x		x	x			4		x	x	x	
	(3)	x	x	x		x		x	x	x		7					
Meakins	(1)	x	x	x		x			x	x		6	2	x	x	x	4
														x			
Missingham	(1)	x				x		x	x			4	3	x	x	x	6
	(2)	xR	xR	xR					xR	xR		5		x	x		
														x			

area 1 { Meakins, Missingham }

Table 8.3: Middle schools: Contributions made by co-ordinators and key teachers

(a) Co-ordinators

	A	B	C	D	E	F	G	H	J	K	No.
Movehall (1)	x	x	x	x			x				5
Movehall (2)	x	x		x		x	x	x			7
area 2 — Makewell	x	x	x	x	x	x	x	xR	x		10
Measures	x	x	x	x	xR	x	x				7(8)

(b) Key teachers

	No.	L	M	N
Movehall (1)	2	x	x	x
Movehall (2)		x	x	
area 2 — Makewell	4	xx	xx	xx
	3	xx	xx	xx
		xx		
Measures	4	xx	x	x
		xx		

Totals: Movehall 6; Makewell 14; Measures 6

(a) Co-ordinators
A Changed teaching

Standing in school
B Head confident in her
C Colleagues respect her
D Colleagues accept her advice

Meetings
E With staff
F With parents
G Contribution to teachers

Preparation of mathematics scheme
H Has mathematics knowledge needed
J Prepared by head and co-ordinator
K Prepared by all the staff

(b) Key teachers
Number
L Changed teaching
M Influenced colleagues
N Acquired more mathematics

R After 1978

heads (Table 8.1), the co-ordinators and key teachers (Tables 8.2 and 8.3) at each project school.

Factors affecting the contributions made by the heads

(**A**) The extent of the mathematical knowledge possessed by the head determined whether or not the heads would be able to take an active part in the project. Although the attitudes to mathematics of the six heads of middle schools during their education were by no means all positive (see Table 2.1, page 00), they seemed to have acquired further knowledge of the subject after leaving college (by reading or by attending courses). On the other hand only two of the first school heads had an adequate knowledge of mathematics. Both used their knowledge to give sustained help to their teachers, since the co-ordinators could not do this.

The extent of the overt support given by the heads to their teachers was indicated by: (**B**) the way in which they co-operated in the project; (**C**) the amount of informal help they gave; (**D**) the help they gave by working with teachers in their classrooms; and (**E**) the teaching they each fitted in on a regular basis.

The extent of the support given by the heads to their mathematics co-ordinators was shown by: (**F**) the training they themselves gave to the co-ordinator; (**G**) the provision they made for the co-ordinators to have non-teaching time to inform themselves about standards and about where help was needed (and to provide this help); (**H**) the provision they made for staff meetings (for discussions, preparation of schemes, workshops) to be taken by the co-ordinators for the entire staff or year groups; (**J**) the encouragement they gave to the co-ordinators and key teachers to arrange a working session for the parents. Such working sessions had an additional spin-off, as the extensive preparation necessary served as in-service education for the teachers. Moreover, the LEA asked the heads of Fleet and Makewell to arrange working sessions in mathematics for parents throughout the borough.

The extent of the support given by the heads to the project was also indicated by: (**K**) their attendance at the centre-based working sessions; (**L**) their attendance at my regular sessions with able and with slow-learning children. Because the heads were able to take an active part in the subsequent discussions in the staffroom (which I organized after every session) the teachers were made aware of the head's interest; this indirectly supported the work of the co-ordinator.

(**M**) One of the most influential factors in promoting and facilitating the changes which were taking place in the teaching of mathematics was the part played by the heads (and the co-ordinators) in the preparation and subsequent trial of a mathematics scheme. By custom, the heads of primary schools had formerly taken the responsibility for all the schemes of work in their schools. At the start of the project the only school with a current mathematics scheme (prepared by the new head) was Foster. Early in 1978 the head had asked the staff to appraise the scheme. At my next visit she told me that 'The scheme was pulled to pieces and rewritten, but I am relieved that the basic thinking is the same.' Sections had been added on mathematical language patterns and on the practical assessments which all the teachers were using. From then on, the staff co-operated fully in adopting the scheme because they had been actively involved in appraisal and revision. Some heads assigned the task of making a scheme to the new co-ordinator, who discussed the draft with the head before presenting it to the teachers (Finlay, Meakins and Measures). Other heads co-operated with the co-ordinators themselves in this preparation (Frame, Flanders, Movehall and Makewell). But the heads who proved the most successful in the subsequent implementation of the scheme were those who involved not only the co-ordinator but all the teachers (Fleet, Foster, Movehall and Makewell). Among these the procedure evolved by Fleet was outstandingly successful.

To summarize, the totals in Table 8.1 show that the heads of Fleet, Foster and Makewell contributed in every possible way to the implementation of the project. The head of Measures also contributed in many ways. But the heads in three of the first schools made very little contribution to the project during the first two years.

Factors affecting the contributions made by the co-ordinators

At the start of the project neither the head nor the co-ordinator had been briefed by the LEA about the role of the co-ordinator. There was therefore a good deal of uncertainty and anxiety among the co-ordinators, particularly those from the first schools.

Not surprisingly all had begun by reviewing the mathematics equipment available within the school. (As a result of the working sessions, many ordered new material later on.) None began on one of their major tasks – the preparation of a scheme for mathematics – since that would have required self-confidence, professional expertise in the teaching of mathematics and a knowledge of the subject which only two of the co-ordinators (at Movehall and Makewell) possessed at that time.

(**A**) The next task of the co-ordinator was to ensure that the mathematics teaching in her own classroom reflected the changes in the rest of the school which she intended to bring about. Such a classroom would provide an example for other teachers. Furthermore, her experience of effecting the necessary ch.nges in her own classroom would enable her to talk confidently to her colleagues about the problems she had encountered, and her failures as well as her successes. Only two of the co-ordinators were already providing the children in their classes with structured activities and opportunities for discussion.

I therefore decided that at the initial support visits I would begin by helping the co-ordinators themselves to teach mathematics in an active and questioning way. Not all of them were willing to make these changes, particularly when they lacked confidence in their own knowledge of mathematics. Some learned more mathematics at the working sessions, and subsequently by reading or by attending other courses.

Having a mathematics co-ordinator with status within the school was as crucial to improvement in the teaching of mathematics as having a supportive head. The head's functions were those of facilitator and exemplar. She, of course, had status by virtue of her position; the co-ordinator had to work hard to attain her status. Her standing with her colleagues depended on: (**B**) the confidence the head placed in her; (**C**) the respect her colleagues felt for her professional expertise and knowledge of mathematics; (**D**) their recognition of her as a colleague to whom they willingly turned for advice. Certainly imaginative teaching and a good mathematical background were not enough by themselves to give status to a co-ordinator (Meakins).

Even when heads had made the appointments themselves, they did not automatically have confidence in the mathematics co-ordinator, since when they made the initial appointments they had not been informed about the nature of the role. Moreover, even when the co-ordinators knew what the LEA expected of them, it still took time for them to establish themselves in their new roles. Some of them left without ever achieving status. By the end of the first two years of the project all but one of the original five co-ordinators at first schools had left. Three of the new co-ordinators had missed all the working sessions; the fourth had attended as a key teacher. At the middle schools, the changes made in teaching style proceeded at a faster pace, perhaps because of the co-ordinators' more secure knowledge of mathematics. Yet these schools were also suffering from problems of staff turnover: there were two changes of co-ordinator. In both phases, those appointed to succeed the first co-ordinators had the advantage of knowing in advance what they were expected to do.

(E, F, G)   Once the heads had accepted the LEA's job description for the co-ordinators, they were persuaded to provide them with non-teaching time. The co-ordinators could then inform themselves about standards and give help and encouragement where they were needed.

(H, J, K)   Eventually all the co-ordinators became involved in preparing and trying out new schemes for mathematics.

It will be seen from Table 8.2 (see page 000) that during the first two years of the project no significant contribution was made by any of the co-ordinators at first schools. (All of them except one had made substantial contributions by the end of the project.)

At the middle schools (see Table 8.3, page 000) one co-ordinator (at Makewell) contributed in every possible way from the beginning of her appointment. The co-ordinators at Melia and Measures made contributions in several respects. (By the end of the project all the co-ordinators of middle schools except the one at Missingham – recently appointed – had made substantial contributions to change in the teaching of mathematics within their own schools.)

The rather slow start made by some of the co-ordinators is not surprising in view of their belated briefing and of their attitudes to mathematics during their own education. For example, the second co-ordinator at Movehall said at her first interview:

> I was hopeless at mathematics at the secondary school. I just scraped O level. My exam result was achieved after a real struggle even with extra coaching . . . I went to a college where the course took place during the first year, for two hours a week. The lectures

were from 6pm to 8pm on Mondays. Since then mathematics has been my dread.

At the second interview she said:

> I now have a quite different attitude to mathematics. I have really enjoyed teaching mathematics. The project has helped in this . . . I feel quite good. I have enjoyed giving the first and second years topics like the area of hands and feet which they could present attractively.

This young teacher, in her first post, had worked hard at preparing materials for the children to use. Despite her anxiety about mathematics she gained confidence in a short period, mainly because of the increased interest shown by the children.

The co-ordinator at Makewell, who received the full co-operation of the head, achieved a greater change in the teaching of mathematics in the school than any other co-ordinator, despite a high staff turnover (70 per cent by 1978, 85 per cent by 1979). This achievement was of particular interest because when she left school herself, her attitude to mathematics had been negative. At her first interview she said:

> At school mathematics was very formal and was taught as three separate subjects. I disliked maths intensely – but nevertheless managed to pass O level. The light began to dawn at college. The approach, for us as well as the children we were going to teach, was, 'Do and understand'.

Of her teaching she wrote:

> I rely on my own experience and knowledge first – teaching a topic or concept in groups – backed up by a selection of appropriate school maths textbooks.

At the second interview her main concern was to show me the progress made by some of her colleagues. The only problem this co-ordinator did not resolve was that of very experienced colleagues who were reluctant to make changes. She enlisted my help with these teachers. The teachers at this school told me how much they appreciated the encouragement the co-ordinator gave them for every effort they made, and her readiness to use their work to help less confident teachers. Was she sensitive to her colleagues' difficulties because of her own experience when at school?

When there was a change of co-ordinator during the first two years of the project, the factors applying to each are shown separately in Tables 8.2 and 8.3, which also show the extent to which second-phase co-ordinators were beginning to carry out their tasks as described by the LEA. However, not all of the changes of co-ordinator at the middle schools were for the better; some were inexperienced when the assessments were made.

Factors affecting the contributions made by the key teachers (and all teachers at on-site schools)

Seven key teachers out of a total of twenty-six made rapid changes in their own teaching styles. Like the co-ordinators, key teachers could not influence their colleagues until they had experimented themselves. They had to become convinced that the provision of activities and the creation of opportunities for discussion about these activities were important concomitants of the successful teaching of mathematics. All these seven key teachers influenced other colleagues by their own example in their classrooms; in other words, they functioned as key teachers according to my definition of this role. An equal number made gradual changes in their own teaching of mathematics but had no influence on their colleagues.

In all the project schools except Fowler the key teams were asked by their heads to recount, either to them or to all the teachers, what had happened at the working session on the previous day. But it was the key teachers' personal discussions with their colleagues (usually those teaching children of the same age group) which were more effective in keeping them informed of new possibilities. It was their own example in their classrooms and their subsequent discussions of children's actual responses to new activities and games which encouraged other teachers to begin to experiment. Usually these other teachers also asked for my help during the support visits.

There were various ways in which the key teachers (and the co-ordinators) changed their teaching and set good examples for their colleagues. Perhaps the most important, and the least difficult, in that it did not require a change of organization, was using an encouraging manner with the children. I had strongly urged the key teams to avoid saying 'That's wrong' to children but instead to ask them to describe how they carried out an activity or a calculation. (Usually, while talking about their work, children come across their 'mistake' with little prompting from the teacher.)

Another change made by key teachers was the organization of 'talking mathematics' sessions. Key teachers began to ask individual children in a group to talk about a method they had used for a mental or written calculation or how they would tackle a problem and arrive at a solution. The provision of 'talking mathematics' sessions, or of other opportunities for peer-group discussion, was by no means easy, even for experienced teachers, more especially if they were at rather traditional schools. At an early support visit a first school teacher near retirement said:

> I find it difficult to organize talking sessions. I have tried but it is not easy with a class. Talking with a group – there are always interruptions from the rest of the class.

By the fourth support visit three months later this teacher had solved the problem by organizing her fourth year class informally; they were working in groups on different aspects of a chosen project. In consequence the children were far less dependent on her. Younger teachers took longer to introduce talking sessions, especially if they had problems in controlling the children.

A third type of change observed in the teaching was in the provision of on-going activities in which the questions asked by the teacher helped the children to acquire a concept or solve a problem but did not tell them directly. This change required a fundamental shift of emphasis in teaching style and took some time to implement. It involved the organization of group work with more than one group engaged on activities at a time. It required careful planning, the provision of equipment and, above all, training the children to accept responsibility for working in this way. Even experienced teachers with good class control encountered problems when making this change. Those teachers who were unsure of their control required help from the head or the co-ordinator, or from me at support visits, in order to make even a gradual change. Those schools temporarily without a co-ordinator (Frame, Missingham and Movehall) to advise and give encouragement were at a decided disadvantage.

The majority of the teachers at first schools were accustomed to planning work for groups of children. Except for inexperienced teachers, the change in teaching style was not so great at this phase. But a young key teacher who relied almost completely on workbooks for her six-year-olds took longer to convince. It was more than two years before I could persuade her to provide some activities for the children. Ultimately, to her surprise, this teacher became independent of the workbooks. At middle schools, where there was far more class teaching at the beginning of the project, the change of teaching style, even for some of the key teachers, often took longer than at first schools. Two young teachers at Missingham, one a key teacher, achieved the change by organizing team-teaching. They shared the classes, one having a group much smaller than the other for mathematics. Both teachers thus became confident at organizing group activities.

Most middle school teachers approached this kind of change cautiously. One experienced teacher (at Measures) began by preferring class teaching (and therefore instruction). At one support visit he asked me to take subtraction with his class. After discussing possible activities I asked him to organize the class in groups. As he moved from group to group he was astonished to find the number of children (nine-year-olds) who used their fingers to find the difference between the scores on two dice. He found this disturbing. Finally, when he began teaching a slower set of second year children, he became convinced of the value of group activities, because this organization enabled him to observe, without giving the children a written test, how they carried out their calculations.

One important function of key teachers was to support the co-ordinator at meetings with the head, the staff or the parents. Such support gave the co-ordinator and the key teachers more confidence. There were two schools (Melia and Makewell) at which the co-ordinator and key teachers formed a strong team with a common purpose. This increased their joint contributions. At both schools the teams' discussions in the staffroom, and their individual examples in their classrooms, caused several of their colleagues to attend the LEA mathematics courses. Furthermore, at Makewell, where the head was a strong contributor to the work of the key team, when key teachers left the school on promotion, replacements were easy to find because so much enthusiasm and knowledge had been generated.

When any teacher seemed to come to the end of her resources in mathematics one of the team would come to her rescue.

Another change observed in some of the key teachers at several first and middle schools was their determined efforts to increase their own knowledge of mathematics. Fifty per cent of these teachers improved their mathematical background by attending courses or by reading. They all gained confidence as a result.

Seven key teachers (one from a first school and six from middle schools) left their schools during the first two years of the project on promotion to deputy head or to mathematics co-ordinator. Another seven left during the first three years on maternity leave.

Tables 8.2 and 8.3 give an indication of the relative effectiveness of individual key teachers according to the three factors: change of teaching style (**L**); influence on other colleagues (**M**); increase in their own mathematical background (**N**). (As before, these factors do not have equal weight.) The contribution of the key teachers at Makewell was more than twice that of nearly every other school (in terms of individual factors). Among the first schools the contribution of the key team at Fleet was the highest. The contributions of the key teachers in the three first schools in the area of social priority (area 2) were all low. More key teachers in the middle schools than in the first schools influenced their colleagues and increased their own mathematical background.

CHAPTER 9

Developments during 1979 and 1980 and a final comparison of the total input and the estimated change in the teaching of mathematics in the schools

Introduction

During 1979 and 1980 there were two components which helped me further in my assessment of the project: one was unexpected, the other was part of the project design. The first was a week's conference on 'Transition Years 7 to 9 Mathematics', organized by the mathematics adviser. The second component was the series of regular visits I paid to each project school to discover whether, in the absence of any further input, changes in the teaching of mathematics were being maintained. (In the course of these visits I continued to give support where it was needed.) This check was particularly important in view of the high staff turnover at many of the schools.

Mathematics conference on the transition from first to middle schools

Early in 1979 the mathematics adviser invited me to help her plan a conference on transition. This adviser, who had been following the progress of the project closely, planned to utilize the expertise of the heads and teachers from some of the project schools. She suggested that three heads and seven co-ordinators from project schools should be invited to contribute to or participate in the conference. Two heads from first schools were asked to lead discussions, one (from Foster) on 'Assessment and Recording', the other (from Fleet) on 'Making and trying out a Scheme'.

The conference provided me with an invaluable opportunity to assess the extent of the changes which had taken place in the attitudes of these heads and teachers to the teaching of mathematics, and the strength of their commitment. Those who contributed had to clarify certain issues for themselves in order to be able to convey these to their colleagues. The head (of Foster) who led the discussion on assessment and recording brought a group of children with whom conference members could try the assessments for themselves, since all the teachers in her school now carried these out as part of their normal programme.

The head of Fleet described her experience with the teachers in her school during the preparation and trial of a mathematics scheme beginning with number. This process had necessitated the crystallization of aims, much reading and frequent consultations, the production of workcards and games, and the preparation of a checklist and a number readiness test. After I had been asked to appraise the draft it was tried out in the classrooms. At the end of the trials I was asked not only to appraise the children's work on display but also to run a workshop which would help the teachers to appreciate the mathematical purpose of the activities they had included.

Copies of the school scheme for 'Area' were distributed as examples to the small groups of conference members formed to prepare schemes on the topics teachers found difficult. This session enabled me to listen to the contributions of the project teachers; these were imaginative and soundly based.

The session led by the co-ordinator (at Fowler) who had prepared and tried out many games to help children to memorize number facts was so successful that she was asked to put on an exhibition at the teachers' centre for a wider audience.

The conference gave me the opportunity to see heads and teachers from project schools using the ideas they had learned and put into practice as a result of the project. In addition, I appreciated the favourable reactions of the other heads and teachers to the active method of teaching mathematics which was being considered throughout the conference.

My visits to individual schools

Background

Throughout 1978 I had continued the support visits, two each term to each project school. Most of the co-ordinators and key teachers had made noticeable changes in their teaching of mathematics. In consequence I was able to concentrate on helping other teachers nominated by the co-ordinators: those in their first posts and others new to the school who required help in implementing the new school scheme, and experienced teachers to whom the co-ordinators were reluctant to offer help. At this stage, whenever possible, I encouraged the teachers to take the initiative while I acted as their assistant during classroom support.

From the beginning of 1979 my visits to schools were not solely concerned with support. I now visited each school once a term with several aims in mind. First, and most importantly, I wanted to monitor the changes which had already taken place. Would these be maintained or even developed further when there was no fresh input? What would be the effect of the high staff turnover in some schools? How would new teachers be helped to implement the mathematics schemes which some schools had evolved? Secondly, I wanted to keep myself informed about staff changes. Thirdly, I intended to ask the heads to assess the extent of change in the teaching of mathematics made by every teacher who had been at the school for more than a year since the beginning of the project. Most of the heads did this in consultation with the mathematics co-ordinators.

I had made my own assessments during my support visits and interviews, based on the following factors.

1. The contribution of the head, the co-ordinator and the key teachers; the preparation and trial of a scheme.

2. The contribution made by individual teachers.
    (a) The nature and extent of the activities provided.
    (b) The opportunities provided for discussion.
    (c) The questioning: does this tell the answer or promote learning?
    (d) Does the teacher interact with the children in a positive way?
    (e) To what extent does she rely on a textbook or workcard system?

3. Confidence, knowledge and understanding (teachers and children).

    (a) Was the teacher confident in what she was doing?
    (b) Did she understand the mathematical purpose of the activity?
    (c) Did the children enjoy mathematics?
    (d) Did they understand what they were doing?
    (e) Were they able to talk about the activity or calculation?
    (f) Did they have sufficient number knowledge to carry out calculations?
    (g) Were they allowed to develop and discuss more than one method?
    (h) Were able children given any special attention? Slow learners?

I was able to compare my assessments with those of the head, the co-ordinator and the teacher herself, and sometimes with those of the advisers. When the head had made her final assessment of each teacher I discussed my own assessment with her. The results were subsequently expressed as percentages: that is, the estimated extent of change in the teaching of mathematics made by the teachers at that school since the beginning of the project. When there was a discrepancy between our estimates I took the lower one.

The accounts of individual schools which follow cover the development of the project between January, 1979, to August, 1980.

First schools: Frame, Fleet and Foster (area 1)

Frame (centre-based pattern of working sessions)

This school continued to have a low staff turnover (40 per cent); the three key teachers remained at the school until the end of the project.

The head's traditional philosophy was modified to some extent, although her response to the 'back to basics' campaign was to insist on the rote learning of the multiplication tables by the fourth year children. After I had compared my own assessments of the changes made by individual teachers with those of the head she said:

> I should not have been able to discuss and appraise so frankly three or four years ago. I was always on the defensive then. Now I am

more relaxed. Attitudes have changed in the staffroom. All the teachers are now willing to talk about their problems and failures. They use more material in their classrooms. I'm sure the project has made a lasting difference. People are not conscious that they are doing anything different because they aren't thinking about the project.

The co-ordinator also made shrewd comments about the changes effected by her colleagues. Of herself she said:

I look more at what the children are doing and use this as my starting point. I look for new ways of doing things all the time. The project has opened my eyes. I'm developing more ideas now – but I'm worried about the [new] teachers in the first years. I've no time to visit them but I've suggested that they leave written work for the children until the last possible moment. The longer they leave this, the better.

The comments made by both the head and the co-ordinator showed significant changes of attitude to the teaching of mathematics. The head recognized the change in the co-ordinator, too. Shortly after my visit she gave her time to help the new teachers in their classrooms. The co-ordinator also told me that she was reading to increase her mathematical background and had applied to attend an extensive college course for co-ordinators. She admitted that she was far more confident and that she welcomed me as someone with whom she could discuss her problems.

The key teacher who had made a dramatic change in her teaching style in every aspect of the curriculum supported this view of the changes made by the teachers:

The whole staff has changed – they discuss their problems. The course and your visits have opened their minds . . . Your support visits were useful, especially seeing how you dealt with the children.

These comments made me wonder whether my estimate of the overall change made by the teachers (40 per cent) was too low; the head's estimate was 60 per cent, but I realized that her expectations were lower than mine. Moreover, one senior teacher who had made many changes herself said:

The project has not had much effect on the staff as a whole. Teachers need to get together and thrash out the scheme. We've had two discussions on it so far . . . [The project] made me think about maths in a different way. I've made workcards. Games were useful, too. I made copies and sent to a friend, a head in another borough.

Discussion with this teacher tilted the balance; I decided to adhere to my estimate of 40 per cent as the total change in the teaching of mathematics.

Fleet (centre-based pattern of working sessions)

Although the staff turnover throughout the project remained at approximately the same level (a total of 60 per cent by the end of the first three years and one term) the changes in teaching styles continued without interruption until the second co-ordinator left on maternity leave and could not be replaced. Another key teacher, the deputy head, left at the same time. The two factors which had originally caused the changes to gather momentum were the active involvement of the head and the gradual preparation of the mathematics scheme. When the scheme was complete the head became critical of the first stage, the number section. She said:

If I started again I would supply and feed in practical ideas for the number scheme. The present scheme is far too complex. Enormous emphasis would now be placed on 'talking mathematics'. Things we think the children know, we've found they don't. We are horrified at the number now. We wanted to have everything on paper at that time. Now we would include much more about talking. We have to work hard with the parents to satisfy them, too.

This comment shows the significant changes which had taken place since the beginning of the project in the head's attitude to the teaching of mathematics. She was clear-sighted in her assessment of the reaction of the parents to such a radical change as the shift of emphasis from premature recording to investigation and discussion. She organized sessions for the parents to keep them informed. She appreciated the difficulties some teachers had in effecting changes and gave them encouragement; she helped individual teachers to organize activities and to ask questions which focused attention on important concepts. Although the head made her assessment of the extent of the changes in the teaching of mathematics on different criteria from mine, we both agreed that the overall change was 70 per cent.

When I interviewed some of the teachers their views supported those of the head. The deputy, who left in 1979, made comments which were in marked contrast to those she had made in 1977:

> I don't think you would like the way we are planning the new [number] scheme.

In 1979 she said:

> I think the changes are permanent. The number of people involved in the changes is sufficient to ensure that anyone coming new to the school will be absorbed, especially since the head is involved herself.
>
> I enjoy maths now. I was dabbling before. I know where to go for help now. I feel much more confident. Basically the project has caused this. There is a tremendous improvement in talking. I did not understand the possibilities until we saw you working with children, and worked with children ourselves. We now know the mathematical reason for things.

This teacher had been one of the least confident because of her scanty mathematical background. She had clearly benefitted from the preparation of the mathematics scheme, in which, as one of the key team, she had been heavily involved at all stages.

The second co-ordinator, trained by the head to take this responsibility, also expressed her view of the progress of the project objectively:

> Before the project, teachers did not see each other. Now we have many informal conversations. I think a lot about mathematics now. Even before the head came, the few key teachers made a difference. When they returned from the working sessions with ideas they made me look closely at what I was doing. There are still

weaknesses but changes happen because we all work together. We are a very happy school.

There were two teachers who still required much encouragement to increase the scope of mathematics. Both were made more secure by the scheme they helped to prepare. One said:

> The children seem to enjoy maths with the changed approach. I'm more aware of maths . . . maybe I make more of opportunities which arise. We do more formal work, too . . . The project has opened my eyes. I am more alert and aware. I integrate maths more with other things. I think there is a good attitude to maths in the school. I feel more confident.

The second teacher, who had been very dependent on workbooks, voiced another problem:

> I try to talk more – and have a lot more conversation about maths. I bring more into maths . . . It helps to have one person to feed in ideas. But I find very able children difficult. Most of us resort to giving them a book.

Despite the high staff turnover the teachers at this school made greater changes in their teaching styles than those at any other school. This was the only first school to integrate mathematics into specific projects. These developments resulted, in the main, from the head's initiative and determination once she appreciated the value to the children of the changes proposed.

Foster (on-site pattern of working sessions)

The changes in the teaching of mathematics at Foster developed in a somewhat different (but almost equally successful) way. Until September, 1978, the staff turnover was the lowest (33 per cent) of all the project schools. (Subsequently three teachers left during 1979 and four in July, 1980.)

The head of Foster, like that of Fleet, had been appointed at the time of reorganization in 1974 and was in her first post. She faced different problems. It had been evident from the beginning that the three 'resisters' out of the nine teachers at Foster had a greater influence on the attitude of the other teachers at that school than had the two resisters out of the thirteen teachers at Fleet. Moreover, since the working sessions at Foster were on-site the resistance of these teachers could not be concealed.

Both schools had mathematics co-ordinators (appointed before the heads) who were unable to help their colleagues and who preferred dealing with older children and class teaching. In each school the head therefore took the initiative and actively supported the project until she had trained a co-ordinator of her own choice. The head at Foster had appointed a young teacher from the middle school (who had attended the on-site working sessions at Foster) to be mathematics co-ordinator when the first one left in 1979.

The head at Foster had already prepared a mathematics scheme herself which, later on, was appraised and revised by the whole staff. This was a concentrated, not a lengthy, process. Her major contribution to the in-service education of her teachers was the opportunity she gave them to take practical assignments with two children at a time. In this way the teachers gained insight into the way children learn mathematics by means of structured activities. In consequence they modified their own teaching styles. Gradually all the teachers took this responsibility for assessing the children they taught.

After the parents' evening in which all the teachers had provided for the parents the mathematical activities and games they used with their children, one teacher commented:

> We would not have been able to do this before the project. I should never have had the confidence to talk to the parents as I did, before the project. What a lot of maths we are doing now!

Shortly after this evening the head told me that she had had to tell the teachers 'to lay off mathematics', since this subject was dominating the curriculum.

The training of the new co-ordinator by the head took some time. The head said of her:

> She visits her colleagues to appraise what they are doing but not to advise them. I still hold the end of the rope – and would gradually let go as the co-ordinator becomes more confident.

(The co-ordinator had already begun to increase her mathematical background by studying for the Mathematics Diploma.)

Assessing the changes resulting from the project the head said:

> The greatest change has been the exchange of ideas. Teachers are coming to accept that children can be taught mathematics without telling. The whole outlook is changing – the staff are teaching mathematics with understanding. Now we have the active co-operation of a maths co-ordinator I have appointed myself there should be no backsliding in mathematics.

The deputy head, who had been an early 'resister', assessed the project:

> I think the early resistance was because teachers felt they had been conned! They did not know you and had not realized what you were trying to achieve. I've found some ideas and activities useful ... I'm more aware of language, and of the importance of understanding. I don't find the assessments easy – but they help me to know where I'm going, and therefore help with planning.

The head confirmed that this teacher had really changed her teaching of mathematics, and this was supported by the adviser who visited the school. The head agreed that the two other resisters, both experienced teachers trained to teach juniors, had not changed a great deal. Her total assessment of change in the teaching of mathematics was 75 per cent. My estimate was 65 per cent.

These three first schools were in an area of mixed social classes. All three schools seemed to have made changes commensurate with the heads' knowledge of mathematics, their view of its importance in the whole curriculum, their sensitivity to the anxiety of teachers about making changes in the subject, and their determination to effect improvement. In each school I compared the lower estimate of change (mine or the heads') with the total contribution made by the head and the key team (see Table 9.1, page 000).

The first schools in the area of social priority, now to be considered, made far less progress towards changing the teaching of mathematics.

First schools: Flanders, Fowler and Finlay (area 2)

Flanders (Centre-based pattern of working sessions)

This school had had many setbacks during the project. There was a high staff turnover: a total of 80 per cent for the first three years and 50 per cent in the following year. In all, there were three co-ordinators and there was an interim period when the deputy head accepted responsibility for mathematics. Both the key teachers were in their first posts and were unable to help their colleagues. Until 1979 there was no one on the staff who had sufficient background knowledge of mathematics to provide effective leadership in this subject. The head herself felt inadequate in mathematics and did not offer to help the teachers. She first became aware of the need to change the teaching of mathematics when she observed me working with a group of slow-learning children, aged seven and eight years old, and realized how little they retained of the number knowledge their teachers had tried to teach them for three or four years.

When the first co-ordinator left on promotion, the head had tried to secure a replacement with a knowledge of mathematics, but unfortunately this requirement had been omitted from the advertisement. The new teacher, appointed in 1978, had a special interest in language. Nevertheless, she accepted the post of mathematics co-ordinator and attended an LEA workshop on this subject. When the head asked her to prepare a scheme, they both asked for my help; I readily agreed. During the following year the support visits consisted mainly of working sessions which I organized for the head and the co-ordinator, to help the latter to prepare a scheme with a wider scope than number facts and calculations. We also discussed ways of introducing the scheme to the teachers. It was unfortunate that this co-ordinator, who had worked so hard to finish the scheme, left on maternity leave before she could put the scheme into operation. However, the head herself began to show more confidence in her understanding of mathematics; it had been her pressure which had induced the co-ordinator to complete the scheme before she left.

The third co-ordinator was transferred from a school about to close at which she had already been mathematics co-ordinator. At that school she had successfully introduced a commercial mathematics scheme. At Flanders she was hesitant about introducing the previous co-ordinator's scheme. Instead, she began by asking the teachers to outline their expectations in mathematics for the children they taught. She also ensured that her own classroom (for reception children) reflected her ideas for teaching mathematics. She was imaginative in her use of other aspects of the curriculum for mathematics and her work was soundly based.

While the original scheme was being prepared I had failed to persuade the head to arrange a mathematics evening to inform the parents of pending changes. She now began to take the initiative. As a start she organized an exhibition to show the range of topics covered by mathematics; there was an emphasis on the development of language throughout the subject. The exhibition included a display of children's work on 'Time' to illustrate the progression of this topic through the school. (As usual in this school much of the display showed the firm direction of the teachers. But the display was a step forward.) However, the head expressed disappointment that the new co-ordinator came to consult her so often. I discussed this problem with the mathematics adviser who had known the co-ordinator at her former school. She confirmed my views by commenting that 'This mathematics co-ordinator has good ideas, but she would think that the head likes to be consulted.'

During 1979, for the first time, I was able to discuss with the head the changes made by individual teachers since the beginning of the project. (This was additional evidence of the head's increased confidence in mathematics.) We agreed on an overall change of 35 per cent, despite the high staff turnover. We both hoped that with the new co-ordinator the rate of change would increase. But the future of the school was in the balance, since, as a result of falling rolls, there was the possibility of amalgamation with the middle school.

Fowler (centre-based pattern of working sessions)

This school had suffered disadvantages from the outset. At the beginning of the project the buildings were overcrowded; access to huts was across the middle school playground; there was often friction between the two schools. There had been a high staff turnover for many years. The head's meetings with the staff had become opportunities for her to impart information rather than promote discussion.

At first there was no one on the staff with a confident knowledge of mathematics or with any interest in the subject. The first co-ordinator and the two key teachers left after the first input of the project. As soon as the second co-ordinator was appointed she co-operated with me to the full. She always took advantage of the support visits to ask my advice about the scheme she was preparing. She quickly gained confidence and changed her attitude to the teaching of mathematics. She no longer believed that the understanding of concepts was unnecessary. There were several reasons for the changes.

1. Her increased knowledge and understanding of mathematics.
2. The head's appreciation of her worth, which she showed by allocating time for her to visit her colleagues in their classrooms.
3. Her success with the activities and games she used in her class.
4. The co-operation she received from senior colleagues when she discussed the draft scheme with them.
5. The response of the teachers to the scheme and their willingness to accept informal help from her.
6. The appreciation of other teachers at the 'Transition' conference for the games she had tried with her children.
7. The success she had when including mathematics within an integrated topic which she introduced to her class.

Although the co-ordinator tried to secure more support from the head for the project by emphasizing the benefit individual teachers could derive from my visits, the head remained anxious lest I should upset her teachers, probably because the head of the middle school, who had worked with me in the past, shared this fear initially. After a year of increased absence herself, the head retired early and the deputy, also near retirement, because acting head. The co-ordinator seldom had the opportunity to help her colleagues in their classrooms. However, by now the senior staff were organizing their own staff discussions for the first time. When they asked me to work with them I arranged the session in the co-ordinator's classroom where her collection of games and activities was available, in order to show maximum support for all she was doing. (Two 'resisters' ceased their opposition after this meeting.)

An experienced head was appointed in September, 1979. I visited the school to try to ensure that the co-ordinator was able to carry out her responsibilities for mathematics. The head revealed that she herself had disliked mathematics at school, but she agreed to allocate an hour each week for the co-ordinator to visit her colleagues in their classrooms. I found that the co-ordinator was trying out a battery of practical assignments she had prepared as a result of the 'Transition' conference, to assess pairs of children in different parts of the school. She planned to continue her assessments. But once again, many staff absences interfered with this arrangement. The head herself was often absent and the co-ordinator had few opportunities to carry out her plans.

The co-ordinator's change of teaching style was confirmed by the mathematics adviser. But in this school there was no one with whom I could discuss the extent of the changes in the teaching of mathematics, because the second co-ordinator had not been involved during the first input of the project. The whole of the key team had left with nothing achieved. I assessed the change in the teaching of mathematics as 20 per cent. The potential of the co-ordinator is high; she needs only to be given the chance to carry out her responsibilities.

Finlay (on-site pattern of working sessions)

This school had on-site working sessions. All the teachers were therefore conversant with the aims of the project. But the high staff turnover – over 90 per cent during three years and one term – had constituted a setback. During the project there was a change of head, and a change of co-ordinator; then from September, 1979, there was no co-ordinator. All the key teachers had left by the end of the project. Although the first head (who retired in August, 1978) was fully co-operative, her own limited mathematical background had prevented her from giving active support to the teachers. The new head said she had been happy in mathematics to O level, but could remember nothing about her professional course at college. She had, however, been successful in changing the teaching of mathematics at her previous school. Moreover, she encouraged the co-ordinator to prepare a scheme for mathematics and to begin to help his colleagues to implement this in their classrooms. Hitherto, although he had made substantial changes in his own teaching of mathematics, he had not seemed able to influence his colleagues. However, he left, on promotion, before he had made much progress. When he could not be replaced (because of falling rolls) the head said: 'We are struggling to implement the scheme he prepared.' She was attending a reading course at that time in which she involved all the teachers. It seemed that changes in mathematics were in abeyance and the head asked me to postpone my visits. During 1980 the head had a serious illness and was absent for nearly a term.

When I visited the school in June, 1980, I saw each teacher who had originally taken part in the project at work in her classroom, and had a discussion with the entire staff. I became aware, as I visited the classrooms, that the head's language course had increased the amount of language the teachers were using in mathematics. At the meeting I asked those teachers who had been involved in the project whether there had been any lasting effect. A teacher who had left the school for two years and then returned said:

> The project has had a lasting influence on us. It caused the teachers to introduce more language in association with mathematics and therefore more practical activities. [The former co-ordinator] and I worked together and were influenced by the college of education course. The project working sessions were a continuation of this. I could appreciate what you were doing – I've put more thought into mathematics and extended my ideas . . . You get a greater benefit from this way of presenting maths. But the expectations of the parents worry me – to try to explain the different presentation of maths is difficult. I think I now place less emphasis on mechanical sums. I've tried to bring in more practical activities, more talking and do less practice. I've brought in things from outside and told children that these activities were mathematical. I'm confident when teaching mathematics now. This began at college but I could not relate maths ideas to children. I get more talking by asking questions and this gives me insight into children's understanding.

I had been told by the head that she had persuaded the experienced teacher who had resisted any kind of change, and who preferred a quiet class, to co-operate with another teacher. I was interested in this teacher's comments:

> When I was teaching six-year-olds I found the development of subtraction good. Now I'm teaching older children I tend to fall back on what I knew formerly. The [new] approach to tens and

units has probably changed me a bit. Talking? No change really. I've made a set of cards for measuring activities. When doing these the children talk among themselves. But I've always taught maths in this way. You gave us some good ideas – but when you try these with a class you find it too difficult.

I asked this teacher if she ever suggested that the children should tell her how they were working a calculation. She replied: 'Not as much as I would like.' I asked her why she could not manage this since she was a well-organized teacher. She replied: 'I do enjoy language more than mathematics, but there should be more time. There are so many outside pressures.' I felt that my original assessment of this teacher had been confirmed; there had been little change in her teaching of mathematics, although, because of the head's persuasion, she was now taking part in team teaching.

The teacher with most experience at this school, who now had children of ages four to six years old and who had had least confidence in teaching mathematics said:

The project has helped. It opened my eyes. We do more weighing now. We still do sums once a week because we enjoy them. But we do many other things as well. Change takes time because I am so set in my ways. Before the project we always did addition of tens and units. Now not all children are adding up to ten. Definitely there is more talking and more activity. If you had put pressure on me I would not have changed. I would have dug my heels in. You have never done that. I have changed slowly. The changes have increased since the new head came.

I had assessed the extent of the changes in the teaching of mathematics at this school as 40 per cent. I had discussed this earlier with the first head and the co-ordinator, both of whom had arrived at the same percentage. This visit confirmed the assessment, particularly after discussion with the new head about the teaching of number. She agreed with me that the teaching of number had not changed much. Not more than half the time allocation for mathematics was given to number, so that other more practical aspects of mathematics were included. This had resulted in more talking on the part of the children. The head was hoping to appoint a mathematics co-ordinator in the future.

For a first school which was given school-based working sessions the percentage change in the teaching of mathematics was relatively low. Yet the total contribution of the head, the co-ordinator and the key teachers was relatively high – almost as high as in the other first school receiving school-based working sessions (Foster). Was this relatively small amount of change due to the more traditional attitude of the teachers to their professional responsibilities? Or to the high staff turnover? To the inadequate professional courses the teachers had had in mathematics? To a co-ordinator who found it difficult to work with his colleagues? Any or all of these factors may have been responsible.

Summary

The three first schools in the area of social priority (area 2: Flanders, Fowler and Finlay) made far less overall change in the teaching of mathematics than the other first schools in area 1. The character of the area could not have affected this result, because during the project many problem families from area 2 were transferred to area 1. Moreover, with the exception of Finlay, I had heard no adverse comments from the teachers about the children in the area of social priority (area 2). The high contribution made by the heads, the co-ordinators and the key teachers at Fleet and Foster in area 1 would account for the extent of the changes in those schools. In the same way, the low contribution at Fowler would account for the low percentage change in that school. Was a common factor, contributing to the low percentage change in four of the schools, the heads' lack of an adequate mathematical background and their consequent reluctance to take an active part in implementing the project? All the heads of the first schools regarded reading as of first importance and, until the advent of the project, did not appreciate the contribution mathematics could make to the language development of children. One head had thought the subject unimportant until the project changed her views.

The presence of an enthusiastic co-ordinator, whose own classroom set an example of the changes she was trying to help teachers to make, who had standing with her colleagues and was able to work with them, was of first importance. The heads of Fleet and Foster had appointed and trained their own second co-ordinators. The second co-ordinator of Fowler had changed as a result of her own strenuous efforts (reading, attending courses and consulting me), but to date her skills had not been utilized to the full. The co-ordinator of Frame had progressed in the same way, but not to the same extent.

Middle schools

Four of the six middle schools (Melia, Movehall, Makewell and Measures) made changes of 60 per cent or more in their teaching of mathematics. In these four, the extent of the co-operation between the head, the co-ordinator, the key teachers and myself increased during the period from January, 1979, to August, 1980. With one possible exception, it seems that the teaching of mathematics in these four schools will not regress, despite the high staff turnover which had ranged from 50 per cent to 85 per cent by July 1979.

Melia, Meakins and Missingham (area 1)

Melia (centre-based pattern of working sessions)

In Melia the staff turnover during this period was 67 per cent. At my final visit during June, 1980, three teachers were on maternity leave and there were seven supply teachers in the school. Nevertheless, because of the workcard system the head had introduced, he felt confident about the teaching of mathematics in the school. Of the effect of the project he said:

> The fact that you started to come into the school and the key teachers began to go out – started talk about mathematics among the staff. The impact was not direct but attention was drawn to the subject. The workcard system might not have been introduced if it had not been for the project. It [the project] did a lot overall by creating an atmosphere in which maths could progress. Interest in mathematics has not decreased; in no way has the spark been lost.

While realizing the shortcomings of the system he had introduced, the head was now more aware of its advantages. He continued:

People are not so much in need of help. They no longer ask: 'How do I do this?' They have to use equipment. There are no disciplinary troubles. Able children are able to romp ahead, therefore teachers can give more time to slow children. But the cards do not meet all the needs in maths – for example, the children do not know all their number facts. And sometimes, they have no idea what the cards are trying to teach them! A small number of teachers still need intensive help.

I questioned whether the needs of the able children were met by 'romping ahead' with the cards. They, as well as all the other children, required some attention from the teacher, preferably a teacher knowledgeable in mathematics. The head agreed that the able children deserved some special attention in mathematics, but felt that the cards provided greater variety than working on their own from a textbook – the usual solution for able children.

By 1980 there was a third co-ordinator. The head said of him:

He has been successful with a second set in the fourth year. He has already visited the classrooms of the teachers in the first and second years.

The third co-ordinator commented:

But for the project, I would not have changed. I approach topics in a more practical way, a more systematic way. My knowledge is based on work from the project. There is far more talking; I use an oral approach. I look for opportunities to use more language and rely less on textbooks. I would welcome more courses to give me a greater grasp of [mathematical] development. I did not adopt a progressive approach before.

I was particularly interested in these comments made by a teacher who had not been one of the key teachers and whose only contacts with me, apart from the support visits, had been the occasional mini-working sessions arranged after school at the head's request. (He would also have been influenced by his wife, the first co-ordinator at this school.)

Both key teachers left the school during 1979 to become mathematics co-ordinators at other schools. Both had come to terms with the workcard system before they left. One of them said:

The changes I've made are lasting. The project formed the basis for everything. It is the way I work permanently. The children are grouped, therefore there are many opportunities for discussion. When starting a topic, I use my own introduction and then use the workcards for practice. It took me so long to stand on my own feet. My problem was I never learnt how to do things. I had to learn again.

(I thought this teacher's transformation had been very swift.) The other key teacher commented:

I did not have set ways, therefore I was receptive. The project made me think more of what I was doing. It [mathematics] became the most important subject at that time. I've learned how to link one topic with another. I tend not to use textbooks as much as I used to. I used to use them for ideas, now I draw on my own experience.

The head agreed with my estimate of the extent of the change in the teaching of mathematics (60 per cent), despite the high staff turnover (in all, there were three co-ordinators, and all the members of the key team left the school).

I knew that the head would be retiring in 1981. I wondered whether the workcard system would continue to be used critically by the teachers. If they relied too heavily on the system, would they continue to arrange sessions to fill the gaps? To what extent would they be able to organize group work so .that there was sufficient discussion?

Meakins (centre-based pattern of working sessions)

The high staff turnover at this school continued throughout 1979 and 1980. Several vacancies were caused by maternity leave; in one year three scale posts were in abeyance because their holders were on maternity leave. But in the event, these teachers did not return. The total turnover for the first three years and one term of the project was over 65 per cent. In 1980, eleven teachers left (some on maternity leave) and there were six vacant posts in June, 1980. Because of falling rolls in the borough these posts had to be filled by local teachers.

The head's reluctance to help individual teachers lessened during 1979:

> Teachers seem to need more direction. When I give it, I always feel guilty . . . I know I ought to take an active part – there have been so many changes and probationers.

When the rate of staff turnover increased in 1980 to 65 per cent in that one year the head said:

> I intend to take an active part in the work programme of the new teachers; I regard this as a challenge . . . There are now many disturbed children I have to see individually.

It will be interesting to see how this head divides her time between giving help to the many new teachers and interviewing the increasing number of disturbed children. (The children came from a new estate, housing problem families from area 2.)

The staff turnover had not involved the mathematics co-ordinator until he left in 1979, on promotion. He expressed his own attitude to his role:

> I do not want to be on a pedestal. I prefer informal contact with my colleagues.

(A number of teachers said how much they had valued his informal help.) During 1979 he planned to give help in the classroom to many new teachers in their first posts, but, once more, the help did not materialize. He said:

> I now have no class, but I could not work with the teachers as I had hoped because of staff absence.

Perhaps his own doubts about the benefits of group learning had made him hesitant about offering help in the classrooms of his colleagues.

During 1979 the head assessed the project:

> The project really came at a bad time because we had so many staff changes. But it had an effect on the school. You stimulated us and took us along.

At the same time, the head revealed that she had always thought of her school as a control group 'because on-site schools would always have an advantage'. I assured her that this had not proved to be the case so far, and that equal percentages of on-site and off-site schools had made substantial changes in the teaching of mathematics.

Before the co-ordinator left the school he said of the project:

> It has made me aware of ideas. New ideas have to be worked through and adapted to the conditions in which you are. Some things are easier to apply than others. In a way I avoid the things I think will cause chaos. Here, with mixed ability classes, we have special difficulties. Even when we had some setting, the organization often broke down because the extra teacher was not available.

As soon as the head knew that the co-ordinator was leaving she asked him to make a new scheme, because the former one reflected his philosophy but did not include the detailed development of topics. She realized that the school might not be able to appoint a replacement for some time.

A new co-ordinator was appointed in April, 1980. She was an enthusiast for mathematics and had had a good professional course at college. She had little teaching experience, but she set about carrying out her new responsibilities in a way likely to secure the co-operation of her colleagues. She had asked for their views about progression in mathematics. At my suggestion she attended a weekend mathematics conference which included practical applications in the teaching of that subject.

The one key teacher remaining at the school said:

> Of course the project has changed my views. I now know the value of activities.

A young teacher with whom I had worked on a number of support visits was leaving the school on promotion. He had used a school journey as the basis for a great deal of mathematics, which the children had recorded in an attractive way. He expressed an interesting view of the project:

> The project helped me about things which I had heard at college but which did not mean anything at that time. At college the

activities and group work seemed meaningless and dry because we could not re-group (the children) immediately on teaching practice. But the course was very practical and we made things for the classroom which I have since used. Now I have my own set I am able to re-group them, play games and do lots of practical work. I use the workcards but I work in groups more than I did. I've made many cards myself since I can organize my class in groups and make this work. Group work is good if well-structured. We must have guidelines and know what to do. The children love playing games and I find that they remember the number facts involved. The co-ordinator has helped a great deal. I ask his advice about how to introduce new topics.

I've become more and more conscious of the importance of maths language. I find myself perpetually questioning the children – I don't accept any answer. I have some very bright children.

This teacher had been one of the quick reactors to the project although he had not been appointed key teacher until after the second input. His comments underline the importance of ensuring that probationary teachers receive help in their classrooms during their early years in the profession. He had adapted the individual workcard system to the groups of children he organized within his class – thereby achieving his aims to provide opportunities for discussion while activities were in progress.

The head agreed with me that the overall extent of the change in the teaching of mathematics was about 40 per cent, despite the high staff turnover and the presence of two 'resisters'. The school now has the chance of a new beginning, with the head determined to take an active part in helping the teachers, with a new and enthusiastic co-ordinator and a second teacher with an interest in mathematics who will support the co-ordinator to the full.

Missingham (on-site pattern of working sessions)

This school suffered not only from a high staff turnover (nearly 70 per cent) and from the number of young teachers in their first posts, but from much illness, some prolonged, on the part of senior teachers. There had been a change of head soon after the project began and there was no mathematics co-ordinator until mid-1977 when the new deputy head took responsibility for mathematics. During the following year she had a serious accident and was absent from school for a long period.

The head, who had always enjoyed mathematics at school, gave her full support to the project. On one of my visits during 1979 she said she 'was shocked to find the extent of class teaching in the school'. She was well aware of how deprived the many young teachers in their first posts were for want of a mathematics co-ordinator to advise and encourage them. She therefore decided to appoint from within the staff a young teacher without much teaching experience to this post. She asked me to co-operate with her in training this co-ordinator, who would, she said, 'find out and learn'.

The head asked me to join her for regular discussions with this second co-ordinator to help her to work successfully with her colleagues. The co-ordinator had an enthusiasm for mathematics and said: 'I loved mathematics when I was at school.' But she needed to be persuaded to ask her colleagues for their views, rather than to circulate her own opinions without previous discussion. At our first meeting she commented:

> I hear only of complaints from my colleagues. I never hear about what the children can do well.

The head took the opportunity of stressing that all the teachers required encouragement for their efforts (like the co-ordinator herself). She was urged not to be critical of the colleagues she was trying to help, and to make sure that her own classroom embodied the features she wanted to introduce to them.

At my second visit the head reported on the progress the co-ordinator had made:

> She has changed a great deal but is still over-anxious about making changes in a hurry. She is having an effect on the teachers. I've suggested that she should request the help of the remaining key teacher to organize a staff workshop in mathematics and this has been done.

The second co-ordinator had, of course, missed all the working sessions of the project and most of the support visits. Between my visits she had totally reorganized the mathematics equipment. She had arranged year-group meetings with the teachers of the first three years in which they discussed the ground they had covered. She had prepared assessment sheets which she had given to the teachers saying that: 'These are just a guide; please criticize.' She also had plans to help her colleagues to introduce 'friendship grouping' in their classes when providing activities in mathematics. She said:

> I want to introduce lots of games and links with life. Colleagues lack confidence. The children's attitudes need to be changed, too. I've changed.

Indeed this young teacher had made rapid progress. It seemed likely that with her enthusiasm and capacity for hard work, and with the continued support of the head, changes would soon be effected in the teaching of mathematics which I had not been able to bring about during the project.

The head and I together assessed the percentage change in the teaching of mathematics since the beginning of the project as 40 per cent. This seemed low in view of the extensive input (the working sessions were school-based) but not low when the high staff turnover was taken into account.

Movehall, Makewell and Measures (area 2)

Movehall (centre-based pattern of working sessions)

The project began at a difficult time for this school for several reasons.

1. Initially the school shared premises with the first school.

2. There was a protracted move to newly adapted premises.

3. Half of the teachers were in their first posts.

4. The head's philosophy was new to all members of the staff; they had to learn to cope with team teaching and with integrating some aspects of the curriculum (not mathematics). These demands, which were willingly met by most of the teachers, made it impossible, at first, for the two young key teachers to stimulate their colleagues to change their teaching of mathematics.

5. The skill of the mathematics co-ordinator, who was the only senior woman on the staff, could not be utilized to the full for the project, because other calls were made on her time. (At the end of the first input she went on maternity leave and could not immediately be replaced.)

There was a high staff turnover (85 per cent) during the first three years and one term of the project, because the head felt that he should encourage promising young teachers to seek promotion. However, the support visits had an effect on the teaching from the beginning. This effect increased when the school settled in its new premises. Although the organization allowed at most three hours a week for mathematics, opportunities were provided for the children to be taught in groups (without the threat of chaos). The teachers were therefore willing to try structured activities and to encourage the children to discuss what they were doing.

At most of the later support visits the head arranged for me to have meetings with the staff. These sessions usually focused on the applications of mathematics to other aspects of the curriculum, but they were not productive until the head and two of the teachers took part in a weekend conference. This conference on 'Mathematics and Problem Solving' was organized by the Open University. It completely changed the head's attitude to mathematics and he began to take advantage of any incident as a focus for mathematics. (Two examples were a survey of the position and height of school signs, prompted by a fatal road accident in the vicinity of the school, and the study of water flow, occasioned by a burst water pipe.) The head found that more time was being spent on mathematics and more interest generated in the subject among the children. Together the staff prepared new guidelines for core and option work in mathematics and designed a development booklet for recording the progress of individual children. All in all, mathematics teaching received a great impetus from this emphasis on real problem solving.

During one of my visits in 1980 the head commented on the project and his own teaching:

> The project has helped to change the teaching, mainly because of the support visits, the regular contact with someone from outside school who made suggestions which you agreed with . . . There are now many more practical activities and the children talk far more.

I rely less on textbooks . . . but some books include good practical activities.

The two joint co-ordinators took maternity leave four terms after their appointment (and did not return). For a time there was no co-ordinator, but eventually the head appointed one from within the school. Although he had missed all the working sessions of the project this co-ordinator commented:

> The project has had a tremendous effect on my outlook. I used to think that maths teaching could be predetermined – almost by a programme. I've now become convinced that direct teacher-contact and inter-peer exchange are most important. I feel that a lot of practice we used to do was a waste of time. When I saw you a year ago (at a support visit) I think I was ready to change. I give children plenty of time now for talking and for activities . . . It is most important to give teachers confidence in teaching maths. They need sympathetic support.

This co-ordinator had an adequate knowledge of mathematics, was confident in teaching the subject and seemed ready to encourage the teachers.

Overall, the head and I agreed about the extent of the changes made in the teaching of mathematics since the beginning of the project. We assessed this as 65 per cent. In view of the high staff turnover, this was a commendable achievement. The Open University weekend had given a further stimulus to the changes, because it was in harmony with the head's own philosophy.

It was unfortunate that the third co-ordinator left before 1980. Furthermore, the head was appointed to a larger school in September, 1980. The future of this school, with its falling roll, was now in the balance. Of the original teachers, only two remained at the school.

Makewell (centre-based pattern of working sessions)

Like Movehall, this school had a high staff turnover (a total of 70 per cent during the first three years and one term of the project) because the head encouraged his teachers to apply for promotion. By January, 1978, all the original key teachers had left the school or were no longer teaching mathematics. New key teachers in the lower part of the school had been selected by the co-ordinator to help her with the induction of new teachers into the most efficient use of the mathematics scheme. This proved to be good in-service education for the three teachers concerned, particularly for one, recently recruited to teaching, who felt insecure in teaching mathematics. (She made such progress that she was appointed mathematics co-ordinator when the first co-ordinator left the school on promotion in July, 1979.) Two of these teachers attended mathematics courses to increase their own background knowledge. Two key teachers, including the co-ordinator elect, were very successful in developing the mathematical possibilities within the projects they undertook with the children.

Because of this forward planning, despite the high staff turnover, changes in the teaching of mathematics continued. As teachers were appointed, they were trained to use the scheme by the head, the co-ordinator or the key team. The methods employed by the co-ordinator to help individual teachers were informal and unstructured.

During 1979 the head and the co-ordinator began to criticize the scheme they had introduced. They decided that some sections required extension whereas others should be curtailed or omitted altogether. They planned to involve year groups of teachers in discussions which could lead to a scheme tailored to the needs of the school. This development made me feel that the co-ordinator and the key team had in-service education in mathematics so well in hand that my support visits were no longer necessary. However, the co-ordinator urged me to continue my visits, saying:

> Your visits keep teachers up to the mark. You have no idea how much discussion there is in the staffroom about the teaching of mathematics when you are about to visit us.

The mathematics co-ordinator had made an outstanding contribution during the discussions and the preparation of selected topics at the 'Transition 7 to 9 Mathematics' conference (see page 151). Her experience in making contacts with all the high schools to which the pupils were transferred was unique in the borough. She had taken fourth year pupils to the high schools so that they could see the provision for mathematics, meet the mathematics teachers and make an informed choice of high school.

Early in 1979 the head made some illuminating comments to me:

> Before the project began I warned people about your high-powered mathematics courses. I feared you would upset the teachers at your support visits. I want to congratulate you on the low profile you have maintained. None of the teachers has been upset. All have asked for your help at every visit.

This statement was important in view of the criticism made when the research was being planned. The head gave further support to the notion that I was not privileged by my previous experience as HMI when he proceeded, unasked, to make suggestions for a more effective in-service project in mathematics:

> First, the Director of Education should have given an introductory talk building up the project and its leader. Several of the heads of the project schools did not know you. They knew you had retired and wondered whether you were past it and out-of-date. Secondly, there should have been an introductory high-powered three-day course for co-ordinators or for heads. Thirdly, the school support was valuable. The status of the advisers supporting or observing teachers is also important. Only those advisers whose views and assistance are valued should be used. (For example, the mathematics adviser.)

These comments provided a useful perspective on the project. It was helpful to know the probable cause of the protest made by some of the heads at a routine meeting during the first input of the project, when objections were made to releasing key teachers for working sessions at the teachers' centre. The comments also implied that my former position could have been a disadvantage rather than an advantage.

When the second co-ordinator took responsibility for mathematics in September, 1979, she organized fortnightly meetings with each year group of teachers to monitor their reactions to the material they were using for mathematics. The attractive and varied work produced by the pupils for Open Day indicated that the high standards of work and the pupils' interest in mathematics were being maintained. The head said that the co-ordinator had gained in confidence and was 'quietly getting on with helping the five new teachers'.

His assessment of the extent of the changes made in the teaching of mathematics by all the teachers since the beginning of the project was entirely independent of mine. We both suggested 65 per cent change, despite the high staff turnover.

Measures (school-based pattern of working sessions)

From the outset I had maximum co-operation from the head, the co-ordinator and almost all the teachers. It was a decided advantage at this school that all the staff were involved in the working sessions at the school. It was also an advantage that the aims of the project were not new to the head. Moreover, the staff turnover during the first three years and one term of the project was 50 per cent, the lowest percentage turnover of all six middle schools. Nevertheless, a number of senior teachers and all the key teachers except one had left the school by the end of July, 1978. The co-ordinator left the school for one year (1978–9), because of her husband's posting. But she returned to take responsibility for mathematics once more. The head retired at the end of December, 1979, and an acting head was appointed.

Before he left, the head made his final assessment of the extent of the changes which had been made in the teaching of mathematics. These were partly based on his own observations and partly on his discussions with individual teachers. This assessment was made independently; the head and I used different criteria, but our final assessments were both close to 60 per cent. Only one teacher assessed the change she had made as greater than the head's estimate; I agreed with the head.

At my final visit in 1979 I interviewed all the teachers who had been involved since the beginning of the project. Some of their comments follow. An experienced key teacher, who had been very insecure about her teaching of mathematics, had been using the individualized workcard system with her fourth year set for more than a year. She described her impressions of the project:

You jolted me and gave me a lot of confidence. I'm still struggling. I make the children teach me. They know I'm not much good at mathematics. I've accepted that maths is open-ended. It used to be a 'Yes' or 'No' subject. I no longer say, 'Do subtraction – or fractions – in this way.' But I don't draw them out enough. With 26 children I rarely get round.

This teacher had exerted herself to learn more mathematics by working through all the cards. From the beginning she was honest with the children about her lack of knowledge, asking for their help. Although, when the project began, she had successfully grouped the younger children, she had not managed this when using the cards with the older ones. However, there was a good deal of informal discussion and exchange of ideas.

Another experienced teacher commented:

I found the project extremely stimulating – in some ways over-stimulating. I could not at the time put it into practical use – you have to do this yourself. You then think of ways in which they [the ideas] can be used in the classroom. I had a good maths background. Calculus opened a door – it seemed magic. That was pattern! I've adopted, adapted, improved. In the classroom you want to make people aware and more confident. It is like learning a game. The project has changed my mind completely. There must not be inertia – getting stuck in a rut. I have to see that there is a variety of approaches.

One experienced teacher thought she had changed a great deal. (Her classes had always been silent, the children's voices had rarely been heard and the teacher had given full directions.) She wrote:

I've become far more informal. I think I've become more aware of less able children and keep an eye on them far more and work with individual children. When explaining I try to go to the beginning – I try to see if the child has the necessary basic knowledge and lead her on. One is more sympathetic, more aware. Yesterday we measured ourselves (height, arm span, etc.). We stood and talked about being an elegant shape. One is inclined to have a little more fun!

This comment seemed to support my views and those of the head that the changes had been limited. Activities were now provided where before they had been avoided, and this was a step forward; but the emphasis was on explanation by the teacher rather than questioning to help the children to learn, and the pace had become that of the slowest. Would the co-ordinator be able to help this teacher to make further progress?

Another experienced teacher had recently made major changes in his teaching of mathematics. He said:

> The project released me from the idea that you had to teach maths in a certain way. It opened up possibilities and gave me confidence. My own attitude changed – from having had teachers who had made the subject formal and dry. The project gave me a feeling of relish for mathematics which I hope I've passed on. The new workcard system also opened my eyes. Freedom to talk – it never occurred to me before the project that talking maths is very important and that a great deal is talking to individuals about what they have done . . . Your attitude has made a difference to me. Enthusiasm affects teachers and can give a teacher an appetite for the subject and show the possibility of the subject. I cannot look at maths now in the same way. There's an element of magic in it. I've now got it in maths – it tends to come alive!

These comments show that both the working sessions and the support visits influenced this receptive teacher. The working sessions opened up the possibilities of mathematics as an exciting subject at any level. The support visits helped him to change his teaching style from competent class instruction to the provision of group activities and discussion. Even practice sessions became interesting to the children who came under the influence of his enthusiasm.

The second deputy had been appointed after the first input of the project. She had been afraid of teaching mathematics. When she was asked to introduce the workcard system she attended an LEA course run by the advisory mathematics teacher, to help teachers to introduce this scheme on a group basis. As a result she gradually organized the system to operate with small groups of children in order to allow herself more time for discussion with each group. She paid a tribute to the advisory mathematics teacher who had encouraged her to work in this way:

> Very practical things have rubbed off. I would not go back to not using practical activities. In the past I worked very formally. The very practical approach has changed my attitude to maths. The

content of the cards has given me ideas of things to do. The system is working well.

This teacher, outstanding in all other aspects of the curriculum, had made the most of the individual workcard system to improve her own mathematical background and to give her ideas. She had not found the maintenance of group work easy with the workcards (in particular, able children got far ahead on their own) but her new organization provided opportunities for group discussion and was more economical of her time.

I paid a final visit to the school in 1980, when the acting head, who had a special interest in mathematics, was in charge. He planned to free the co-ordinator from a class in September, 1980, in order that she should work with teachers in different parts of the school, helping them to introduce the workcard system which had not yet been adopted by some of the teachers. The co-ordinator was troubled by this suggestion. The acting head, the co-ordinator and I discussed the implications of this plan and the advantages and disadvantages of the workcard system under consideration. The co-ordinator expressed her point of view:

It gives insecure teachers confidence, but it is too directed and does not stretch the able children. It is not suitable for the very slow either. I do not want colleagues to feel compelled by me to use the system. I would like to show them different ways of using the cards. For example, they could develop a topic in their own way, then use the cards, perhaps twice a week, for practice or as an assessment. We need back-up materials, particularly for the able children.

The difference between the acting head's plan to rationalize the mathematics of the school so that all the teachers used the workcard system, and the views of the co-ordinator who wanted more flexibility in the teaching of mathematics, was not resolved at that meeting.

During the project the teaching of mathematics had shown a gradual change. There were several experienced teachers, all of whom changed their teaching styles by varying amounts, a few completely, others not a great deal. It seemed important for this school that the working sessions were on-site; perhaps this was because there were 22 teachers, and it might have taken a long time to encourage such a large staff to take advantage of the support visits. But the positive attitude of the head and his understanding of the aims of the project were influential at the working sessions, as were the attitudes of the key teachers within the school.

The middle schools in area 2 all made at least 60 per cent change in the teaching of mathematics. The two middle schools in area 1 in which the estimated change was 40 per cent (Meakins and Missingham) show promise for further change in future.

The relationship between the total contributions made by the mathematics co-ordinators, the heads and the key teachers, and the estimates of the extent of change made in the teaching of mathematics.

In general, keeping in mind the cumulative staff turnover during the project, the percentage changes were commensurate with the total input made by the heads and the key teams. (My own input need not be considered, as it was the same for each school.)

Only 11 of the 12 first and middle schools taking part in the project are included in the summary because the head of Fowler was absent throughout the period of assessment. Six of the eleven schools had assessments from 60 per cent to 70 per cent. Two, Fleet and Foster, were first schools, and four were Middle schools: Melia, Movehall, Makewell and Measures. The last three schools were all in the area designated as one of social priority (area 2), which seems to indicate that the type of area did not influence the extent of change. Two of the schools, Foster and Measures, had school-based working sessions.

Fleet, with the highest assessment of change (although the head of Foster made a higher estimate), had a high input but also a cumulative staff turnover of 60 per cent. There were two changes of co-ordinator in 1980 but the head's interest and determination enabled the changes to be sustained while she herself trained the third young co-ordinator. Moreover, during this period, a comprehensive mathematics checklist was prepared by the head and all the teachers to use with individual children.

Until 1980 Foster had a low staff turnover of 35 per cent. The contribution made by the co-ordinator increased rapidly when the head was able to appoint and train her own. Two of the original teachers continued their partial resistance to the proposed changes. The head's estimate of the percentage change (75 per cent) was higher than mine of 65 per cent; since the head was frequently in the classrooms her estimate may well have been correct.

The two middle schools with the highest percentage change (65 per cent) were Movehall and Makewell. Neither had school-based working sessions. Both had a high cumulative staff turnover. The input from the project for Movehall was considerably lower than that for Makewell, mainly because the stimulus for the changes at Movehall, although initially generated by the project, was later intensified by the Open University's 'Mathematics and Problem Solving' project. The rate of change in the teaching of mathematics at Makewell was maintained, despite the loss of an outstanding co-ordinator, because she and the head had prepared another teacher to take on her responsibilities, and the training of key teachers continued.

Table 9.1   The total contributions of heads, co-ordinators and key teachers and the estimated percentage change in the teaching of mathematics

		Cumulative % staff turnover		Total contributions made by the head, the co-ordinator and the key teacher		Estimated % change in teaching mathematics	
		1978	1979	by 1978	by 1979	My estimate	Head's estimate
area 1	Frame	40	50	17	18	40	60
	Fleet	60	60	30	30	70	70
	Foster	35	50	15	25	65	75
area 2	Flanders	80	90	8	13	35	35
	Fowler	65	80	11	10	20	–
	Finlay	95	95	16	18	40	40
area 1	Melia	65	65	22	19	60	60
	Meakins	65	90	17	18	40	40
	Missingham	70	70	15	19	40	40
area 2	Movehall	85	100	17	19	65	65
	Makewell	70	85	35	35	65	65
	Measures	50	60	20	22	60	60

Measures, with school-based working sessions, continued its steady rate of change, despite the retirement of a supportive head, because the new acting head was also knowledgeable and interested in mathematics.

The input at the fourth middle school, Melia, began to decline when the members of the key team left on promotion. However, the enthusiasm of the head was undiminished when the system of workcards he had introduced reached the fourth year. He continued his efforts to ensure that the scheme did not preclude actual teaching, group activities and discussion.

Five schools, three first and two middle (Frame, Flanders, Finlay, Meakins and Missingham) had low assessments. But for all of these schools the inputs were similarly related to the assessments, although the cumulative staff turnover was variable. The highest turnover (95 per cent) was at Finlay, with school-based working sessions. At this school there had been no co-ordinator for over a year and the recently appointed head had other changes she wanted to introduce. By contrast, the input at Missingham, which also had school-based working sessions, increased during 1979/80. This increase, and a correspondingly higher rate of change in the teaching of mathematics, seemed likely to continue once the school had acquired an enthusiastic co-ordinator who rapidly increased her knowledge of mathematics and became more confident. The head was taking an active part in her training. The position was similar at Meakins where the head promised, for the first time, to concern herself with the training of her new young teachers and to give every support to a new co-ordinator with a special interest in mathematics.

Flanders also had a high staff turnover and a head who, until after my support visits in 1978, felt unable to give her teachers active help in mathematics. At this school there were, in all, three co-ordinators as well as a period when there was no mathematics co-ordinator. The third was experienced in this position and the rate of change in the teaching, the lowest of the eleven schools, seemed likely to increase if the head took full advantage of the knowledge and interest of this co-ordinator.

The rate of change in the teaching of mathematics at Frame was unlikely to decrease. It might have increased if one of the original key teachers who had made great changes herself had not retired in August, 1980. The first school omitted from the summary (Fowler) has the potential for considerable change under a new deputy and head, and a co-ordinator who has spared no effort either in her private study of mathematics or in the preparation and trial of a mathematics scheme. In September, 1980, the head agreed that she should gradually work with all the children and their teachers in the newly equipped mathematics room, should organize informal work-shop sessions for the teachers and should work once a week with the oldest able children in the school.

Some of the apparent discrepancies between the total contribution and the assessment of the percentage change have already been accounted for. There may be other variables, not least, perhaps, the ethos of the schools as a whole. But this variable has been omitted, not because it was not apparent but because this characteristic can be assessed only by subjective means.

Conclusions and discussion

What conclusions can be drawn from the observations I made between January, 1979, and August, 1980? First and foremost, in all the schools, however high the staff turnover, changes in the teaching of mathematics seemed to be maintained until my final visits in 1980. Secondly, all the heads, even those without an adequate knowledge of mathematics themselves, have come to realize, some for the first time, the importance of having a teacher responsible for mathematics in the school. Recently, they have made thoughtful appointments, looking for a teacher who will have standing with her colleagues and who will work sympathetically with them (or training one to give a lead). Heads have also come to accept the importance of releasing the co-ordinator to visit her colleagues in their classrooms. Thirdly, the comments made by those teachers still remaining at the project schools, more than three years and one term after the first input, were no longer hypercritical or defensive. They showed a more mature judgement. The teachers had had time to consider, to adapt and experiment – and to accept or reject – the proposed changes, according to their experience with many groups of children. Few total 'resisters' remained, although some of the changes made were not great.

The comments made at this stage by the heads and some of the teachers have some points in common. Chief among these was an appreciation of the value of children's talk, discussing their activities and investigations, and comparing the methods used for calculations. Some of the teachers no longer expected children to use the method demonstrated by them, but encouraged the children to develop more than one method. Several teachers also referred to a gain in confidence when teaching mathematics.

There remains the enigma of the difference between first and middle schools in the estimated percentage changes made in the teaching of mathematics. The importance of the active involvement of the head in the project seemed to have been established in the 12 project schools. Was it only a coincidence that the heads of the four first schools in which the extent of the changes was 40 per cent or less had an insufficient mathematical background to take an active part in the implementation of the project?

A comparison of the assessments made by the heads of first schools of their attitudes to mathematics while at school and at college, with those made by the heads of middle schools, does not reveal many differences.

	At school	At college
First schools	1 positive, 3 neutral, 2 negative	2 positive, 2 neutral, 2 negative
Middle schools	3 positive, 2 neutral, 1 negative	2 positive, 2 neutral, 2 negative

The attitude of first school heads to mathematics while at school was rather more negative than those of middle school heads at the same time. There was no overall difference in their attitudes to mathematics while at college. All the heads maintained that they were confident when teaching mathematics, although only two first school heads became actively involved in the project themselves. The schools with these heads were the only two first schools to make appreciable changes in the teaching of mathematics.

Perhaps the difference in the extent of the changes made in the teaching at first and middle schools stems from differences in the professional courses at college? Until 1960, professional courses and the supervision of teaching practice for infant and junior teachers had usually been the province of education lecturers, few of whom had a special interest in mathematics, who gave 'methods' lectures to the students. More often than not these courses were of short duration, especially for students who would subsequently teach young children.

After 1960, however, a major change in the balance of training teachers began to take effect. This change, directed by the Ministry of Education, involved the colleges in a shift of emphasis from training a majority of secondary specialists to training a majority of primary teachers (largely at infant level). As a result, mathematics lecturers, released from much of their specialist teaching, began to taken an interest in ways of improving both the mathematical background of infant and junior teachers and their methods of teaching the subject. But the large majority of these lecturers had not taught primary children. At a mathematics conference in 1959 attended by members of Teachers of Colleges and Departments of Education and Her Majesty's Inspectors, a senior lecturer in mathematics suggested that all mathematics lecturers should go into primary schools to gain much-needed experience in the teaching of mathematics at that phase. The response was enthusiastic. Furthermore, under the programme of expansion, many new lecturers in mathematics were appointed during the next five to ten years and some principals insisted that the newcomers should gain experience of teaching mathematics to primary children. At some colleges a useful partnership was established between mathematics lecturers and education lecturers.

The changes resulting from mathematics lecturers taking responsibility for the professional courses in mathematics for infant (and junior) students did not begin to take effect until 1963. All the heads of the project first and middle schools had completed their training before this date. Perhaps this was why the heads of some first schools thought that mathematics did not matter for young children? All educationists would agree that language skills were the most important for young children to acquire, but there are probably few who would not now give some consideration to mathematics, particularly that arising from activities. Before 1960 the attitude to the teaching of mathematics to infants was more casual.

The students who were trained to teach junior children were not so disadvantaged. Both the education lecturers who gave lectures on 'methods', and the mathematicians who took responsibility subsequently had more knowledge of the mathematical content of the curriculum between the ages of seven and eleven years (influenced by the 11+ examination at that time) than they had of that between the ages of five and seven years. Moreover, the professional courses were more substantial for junior students and efforts were made to help the students to understand the mathematics they would be expected to teach (usually arithmetic). Furthermore, teachers of junior children felt a responsibility to help them to acquit themselves well in the 11+ examination. They often made efforts to improve their own understanding and knowledge of mathematics to this end. (Certainly the heads of the middle schools in the project had done so.) Their counterparts in infant schools, who were focusing attention on helping children to read, had no comparable incentive. When teachers were appointed as heads of first schools they began immediately to help their staff to improve the teaching of reading. They expected them to include mathematics in the curriculum but rarely helped them to improve their teaching of this subject, which was seldom a factor in considering new appointments.

The fact that four of the six heads of first schools did not think that mathematics was an important subject at that phase, whereas the heads of the middle schools considered that the subject was of sufficient importance for them to acquire more knowledge of the subject, perhaps goes some way to explain the differences between the extent of the changes made in the teaching of mathematics at the two phases.

CHAPTER 10

The project in perspective: looking back and forward

The advantages and disadvantages of the research methods used for this project (action research recorded by mean of case studies)

The major aim of action research has been defined by Halsey 1972) as 'to get something done'. Since my main objective was to effect changes in the teaching of mathematics, I decided that this form of research would give me the flexibility I required for the input of the project: the working sessions and the support visits. Moreover, recording by means of case studies would ensure that each school was treated as an individual and complex whole and that the variables would not be limited to those which could be handled within a statistically based project. Although computer analysis has considerably increased the number of variables which can be considered simultaneously, the methods of conventional research do not permit the inclusion of variables which come to light in the course of a project.

The advantages and disadvantages of action research for this project emerged as the project progressed.

Advantages

Flexibility.   During the first input of the project the major advantage of action research became apparent: I did not have to provide each group and each teacher with precisely the same material in the same order. Throughout the working sessions the balance of practical investigations, sequential planning and discussion could be varied at any time according to the reactions of the teachers. In this way individual needs could be met as soon as these arose. However, the overall content of the first input was maintained as originally planned.

No restrictions were placed on the content of the support visits. All the teachers were free to ask me to help them with any topic they chose. The only condition I imposed was that the organization in the classroom during a support visit should allow the children to participate in planned practical activities and that they should be encouraged to talk about these. The teacher and I worked in harness during the session.

The large amount of material amassed.   I collected a wide variety of material, from each of the 12 schools, about every teacher and head. The material was collected through observation (by the advisers and lecturers as well as by myself), through interviews, through the responses to questionnaires and through teachers' comments (oral and written). Frequently the material from one source reinforced that from another. Only rarely was the evidence conflicting. This detailed information could not have been obtained by conventional research methods of data gathering, because opportunities for frequent informal discussions with individuals were unlikely to have been included in the structure.

The extent and quality of the information I obtained by working alongside individual teachers.   When I was working alongside teachers I learned a great deal about the reasons why many teachers were reluctant to commit themselves to making changes in the teaching of mathematics. Their inhibitions usually stemmed from a lack of personal experience of practical activities and investigations as a way of learning mathematics. This meant that they did not know the mathematical reasons for providing a particular sequence of activities, the way the activities might develop, or the questions which they needed to ask in order to help the children to acquire a chosen concept (and later on, to apply it). Above all, most of the teachers confessed to a lack of understanding of the mathematics they were expected to teach. This meant that they were unable either to plan a sequence of activities on their own, or to know where these were leading and how to develop them further. Some of the teachers were also worried lest, in an unfamiliar situation, they would not be able to control the children. Many of these teachers said subsequently that the support visits had enabled them not only to make a start, but to continue working in this way.

By working alongside teachers I was able to give them enough confidence in me for them to describe their problems and suggest what they might do to overcome these. At the same time I was able to judge the progress they were actually making. Once they had observed and worked with one group of children, they became aware of how much they learned about each child's degree of understanding. They then made a more sustained effort and undertook more lengthy sequences of activities because of my promise to continue my visits until they were confident. They began to plan their own activities and to ask for my help with one or two groups at a time. Finally they were ready to accept full responsibility for working in this way with the entire class. By working alongside them, I knew with certainty when this goal had been reached.

The unexpected findings which I made. Because action research provides opportunities for collecting a wide range of data from many different sources, there is great scope for the discovery of unexpected results. Within the present research I was confronted with several discoveries of this kind, of which the following are examples. I was surprised to find:

1. the persisting belief of some heads and teachers that it was more important for children to be able to perform calculations at an early stage than to understand what they were doing – all these were at schools that had agreed to take part in a project to further mathematical understanding;
2. the necessity for a new scheme in mathematics to be democratically prepared and tried out by all members of staff if it was to be wholeheartedly implemented;
3. the necessity for the head to be actively involved in the project if lasting changes were to be made;
4. the necessity for the head to have enough knowledge of mathematics;
5. a lack of clear-cut evidence (despite the more lavish use of my time in schools with on-site working sessions) that the change in the teaching of mathematics in such schools was greater than in schools with centre-based working sessions.

The realization by the teachers that I was interested in them as people, not as statistics. In statistical research an hypothesis is accepted or rejected according to the results of calculations and tabulations, whereas in action research what matters most is the amount of sustained change achieved by individuals involved in the project. In consequence of the different methods used in these two types of research, teachers are apt to feel that in statistical research they are being 'used' and that additional burdens are imposed on them which will be of little benefit to them in the long run. But in action research, since they are made aware from the beginning that they are participants and that they can change the course of events, they appreciate that the work they are undertaking is for their benefit as well as that of the children.

The frankness of the responses made by the heads and the teachers. The heads and the teachers appreciated the fact that their views would be sought throughout the project and that they were making a contribution to the findings of the research. They were frank in their comments to me, often volunteering their opinions without being asked. They soon realized how frequently I modified the programme or procedures in consequence of their comments or suggestions. For my purposes, then, the major advantage of action research was the readiness with which the teachers and the heads expressed their anxieties about their teaching of mathematics at that time and about the changes I was suggesting. Once they had decided to attempt to make substantial changes, they were equally frank about their immediate and future needs. Since we worked in harness, they were free to criticize the outcomes of our joint sessions and to suggest improvements for the following lesson; we were on an equal footing.

The personal challenge of action research. Action research provided exhilarating and constantly varying challenges. The problems which arose during the course of my research (involving more than 300 teachers) were not always those I had anticipated. To solve them I needed all the skills derived from my teaching experience, as well as from my more extensive experience as HMI. Moreover, humility, patience and sympathetic understanding were essential when dealing with the many human problems which arose. Frequently, the heads supplied background knowledge which helped me to see how to respond to a particular teacher.

I had found the fieldwork for my previous research (which had been statistical) exciting at that time. However, the demands it made on me as a person were not nearly as great as those made by this project, because, in the former, the variables were predetermined and it had not involved building new relationships with so many people.

Disadvantages

1. The difficult role of the change-agent.
2. The necessarily long duration of action research if change is to be effected and maintained.
3. The high staff turnover during such a period.
4. The lengthy process of assessment, involving so many people.
5. The time required to write 12 detailed case studies and to codify the material before valid generalizations could be made.

The difficult role of the change-agent. My approach to the role of change-agent was exploratory rather than structured. It was only as I worked that I became aware of all the problems inherent in the complex role of a change-agent undertaking action research. It was essential to establish personal relationships with the teachers, but equally important to maintain objectivity. The same danger arises in establishing relationships with institutions. However, because I was working in 12 schools there was little danger of becoming totally involved with any one school.

When I began the project I was an outsider without status. I acquired status by working with the schools, but at no stage did I have any power. Everything had to be done by persuasion. It was extremely difficult to help teachers to learn for themselves, but essential if they were to internalize this way of working with the children. It was at first hard even to gain acceptance for the idea that I would work with teachers in their classrooms. It seemed to some to be an invasion of their territory. If, by the beginning of the project, the LEA had informed heads and co-ordinators about the co-ordinator's responsibility for visiting and working in the classrooms of other teachers, the difficulty might have been lessened. As time went by, I became much less of an outsider and my influence increased.

There was another role problem. The effect of the severe criticism which the project received on occasion from some teachers, usually during the working sessions, was to diminish my efficiency for the time being. Although I had asked the teachers for critical comments (together with their suggestions for improvement) I sometimes felt depressed until I could see how to modify the programme to meet the criticisms. This was not easy because the recurring complaints from some teachers, that I made impossible demands on them by suggesting that they should adapt the activities to the needs of the children they taught, sometimes wore me down. I was determined not to yield to their pressure to hand out a mathematics scheme, but the alternative of helping them to prepare their own scheme was much harder for both sides.

Perhaps the most formidable and persistent problem of the role of change-agent was the need to reconcile two conflicting tasks. On the one hand, at the working sessions I had to adopt a lively approach and provide stimulating activities for the teachers which would fire their imagination and persuade them to adapt these activities for the children they taught. On the other hand, at all times, but particularly during the support visits, I had to adopt a low profile so that the teachers would never, at any time, feel pressurized. I had to appreciate every effort they made, however small, and to encourage them to continue their efforts.

Furthermore, I was both adviser and assessor. I had to respond to the teachers' requests for help – and also to appraise the extent of the changes they were making. The conflict between these tasks was not, of course, as great as in a situation where decisions about a teacher's career depend on a final assessment, but the teachers were aware that although I was encouraging them, I was also assessing their progress.

The necessarily long duration of action research if change is to be effected and maintained.   If I had not worked in the schools for more than three years it is likely that few of the changes would have been maintained. The process of change and consolidation naturally varied with individual teachers. But change in a school's way of working took at least three years.

By the time of the second support visit I felt I had gained the confidence of several of the co-ordinators and key teachers from schools with the centre-based working sessions, and of many more of the teachers whose working sessions were on-site. Three of the co-ordinators and eight teachers had made substantial changes during the first input. But not all of the teachers were ready to take advantage of help in their classrooms at this stage. Some of the more experienced teachers required a far longer period for observing how I worked with children and how they reacted before they became convinced that the children benefitted from the practical activities we used and from the discussions which took place. The second input and further support visits were essential for these teachers to consolidate their tentative experiments.

In general, the changes in any one school were maintained despite the high staff turnover.

The high staff turnover during such a period.   The cumulative staff turnover was unexpectedly high. Action research in the 12 schools took, in all (including the final visits), more than four years. Within this long period the majority of the original teachers had left the project schools. Because of falling rolls I had anticipated a more stable staff position; I had not allowed for such a high cumulative staff turnover.

The lengthy process of assessment, involving so many people.   Because of the extensive staff turnover, newly appointed teachers were constantly having to be initiated into new ways of working. There was therefore an increase in the total number of teachers who had to be assessed. However, because heads and coordinators worked closely with these teachers, they were able to help me considerably in making final assessments of progress. This was particularly valuable because the advisory team had insufficient time to make the observation visits they had originally promised. But this meant that I had to consult many more people than I had originally envisaged and the whole process of evaluation was necessarily protracted.

The time required to write 12 detailed case studies and to codify the material before valid generalizations could be made.   Most research recorded by case study takes into account one example, or at most two. In order to counter some of the arguments about the impossibility of generalizing from individual cases I had decided to include 12 schools in the project. For each school I recorded each situation in detail and in context. This ensured that all the data obtained were included. I accumulated a large amount of material, very varied in character. When, at the end of the second input, I first attempted to isolate common themes, it proved impossible. I found that I had to write a detailed case study for each school first. Even then, the crystallization of common features proved to be a difficult and lengthy process. But, although at one stage the detail seemed to cloud the issues, it was precisely the accumulation of detailed information which made certain generalizations possible.

A summary of the findings of the project which should be of help to LEA advisers and to other research workers in the field of in-service education

My most positive finding was that the provision of support for teachers in their classrooms, while they were trying to implement changes in their teaching of mathematics, undoubtedly had an effect on the extent of the changes made by each individual teacher, the permanence of these changes and the number of teachers who became involved. The support visits were spent partly in planning with each teacher whose classroom I was to visit, partly in appraising the resulting work with her, but mainly in the classroom, helping the individual teacher to change her teaching of mathematics. As the teacher gained confidence, I gradually withdrew my help until she was able to take responsibility for the class herself when the children were organized in groups. But it takes a long time for many teachers to reach this stage. After three years the extent of the changes across the board ranged from 35 per cent to 70 per cent in individual schools. My experience was comparable in this respect with that of the mathematics advisory teacher. Noticeable changes became evident after three years of contact time with each of his 12 schools. By 1980, it seemed unlikely that the schools in my project would regress. It became clear, too, that working sessions still have an important part to play in in-service education. These sessions showed the need for change and offered a method for bringing this about. They also stimulated a desire for change by providing some activities which the teachers enjoyed and could try in their classrooms. The support visits helped teachers to overcome the problems they met when making these experiments: the fear of losing control of the class, of not knowing what questions to ask to help the children forward without giving them a direct answer, and of not knowing how to develop a topic further.

Another finding was that in any attempt at curriculum change the school must be involved as a whole, if individual teachers were to make real and sustained progress. When the teachers had an avowed common purpose they could talk openly about their experiences, discussing their failures as well as their successes, and could give each other mutual support. Moreover, when the teachers were working together they were less likely to come to the end of their pooled mathematical resources.

At the on-site schools, both the working sessions and the support visits helped to make the teachers aware of their common purpose: to improve the teaching of mathematics. At the centre-based schools only the support visits gradually united the teachers in their efforts. The fact that they did not find such improvement easy to achieve made it imperative that they should discuss their problems in the staffroom. One head after another reported that this was happening and that they regarded such interchanges as a measure of success. At nearly all the schools the project proved to be the first in which all the teachers co-operated to reach a single target.

Action research did not, however, provide a clear-cut answer to one of the questions I originally raised: the relative effectiveness of on-site and off-site working sessions. The distinction between the effects of the different types of working sessions, those held at individual schools and those held at the teachers' centre for key teachers from several schools, may have been blurred by the more numerous and lengthy support visits. The total time spent at working sessions was the equivalent of four working days; the corresponding time for support visits was twelve or thirteen days. (There were additional days for observation and interviews.) The support visits resulted in every teacher at every school becoming involved in the project within four terms.

However, one great advantage of school-based working sessions emerged: the head was present at every session. He was therefore conversant with the aims and content of the project, he himself received in-service education in mathematics, and he could unobtrusively monitor the reactions of individual teachers. Subsequently he was able to advise me about which members of staff would react most favourably to support visits at an early stage. To have the working sessions on-site was also an advantage to the teachers, provided that there was no strong element of resistance in the school. They liked working with all their colleagues and learning which of them had strengths in mathematics and which were anxious about what they were teaching in that subject. The working sessions were the first and only occasion on which they all worked together on one aspect of the curriculum. Because they were working together they were able to contemplate changing their teaching of mathematics at an early stage. Moreover, they knew me, and were willing to have my support in their classrooms.

On the other hand, four schools with the off-site pattern of working sessions achieved a higher percentage change than two of the schools with school-based working sessions. For these schools the temporary disadvantage of having teachers who were initially unwilling to take advantage of the support visits seemed to be outweighed by the advantage of having a head who would be able to give active support to the project by teaching herself, or by helping her teachers to make changes.

Other findings are also important, particularly for those engaged in in-service education. First, for example, it is essential to have more than one input of working sessions, and during the first input, the working sessions should be organized for homogeneous groups (teachers from first schools on *their* own, teachers from middle schools on *their* own). During the second, contact between teachers from both phases is useful. To achieve continuity of content and method joint planning should be organized.

Secondly, effective mathematics co-ordinators are essential if lasting changes are to be made in the teaching of mathematics. The presence of a co-ordinator means that mathematics has a constant advocate in the staff room. When an LEA appoints mathematics co-ordinators these teachers need to be trained before they take up office, so that they fully understand their responsibilities. (The heads, who have to act as facilitators for the co-ordinators, should also be present at some of the training sessions.) The co-ordinator's most important functions seemed to be: 1. to inform herself about the standards of mathematics teaching through the school; 2. to discover where help was needed and to provide this by assisting her colleagues inside as well as outside the classroom; 3. to ensure that her own classroom reflected the changes she hoped her colleagues would make; 4. to help them to prepare a scheme for the school and to try this in their classrooms; 5. to run workshops for her colleagues as necessary.

There is a further requirement if the LEA advisory service is to be effective in preparing mathematics co-ordinators for their responsibilities. The personnel needs to comprise sufficient advisers/inspectors and advisory teachers (as a support team) to be able to tackle the daunting task of first training the co-ordinators and then helping them to support the teachers in their schools. In some areas there have been substantial increases in the advisory personnel for mathematics during the past 12 years. In 1970, in the 166 LEAs, there were only 6 mathematics specialist advisers. In 1980, for the 97 LEAs there were 88 full-time or part-time mathematics advisers or advisory teachers. But advisers have many administrative calls on their time. (One mathematics adviser estimated that she was able to spend only ten per cent of her time on mathematics.) Many specialist advisers spend at least 50 per cent of their time in administrative duties. Advisory (support) staff without administrative responsibilities are clearly necessary if the task of improving mathematics teaching is to be realistically tackled. Many LEAs now appoint advisory teachers on a two-year basis. They are responsible for working in four schools for one day a week throughout the year (or term). The growth of such appointments is shown by the following examples.

1. A metropolitan authority had one mathematics inspector in 1970. By 1981 there were four mathematics inspectors and forty-nine full-time mathematics advisory teachers and consultants. Each advisory teacher looks after four schools at any one time at the request of the heads of the schools. The school allocates some of its funding for this service. The mathematics consultants have a wider responsibility. Both advisory teachers and consultants spend one day each week on their own preparation and training. Mathematics advisory teachers have been appointed in other LEAs also. These appointments are almost always temporary.

2. A large county authority had one mathematics inspector in 1970. Today there are two mathematics inspectors and seven full-time or part-time advisory teachers.

3. On the other hand a large rural authority has had only one mathematics adviser for a number of years. But there are twenty-four advisers with special responsibility for providing

help with the teaching of reading. The results in the reading achievement tests have improved; those in mathematics have not.

Suggestions for further investigation and research

Replication of the present research, modifying some of the variables

1. This research was carried out in one area only, an outer metropolitan borough. Would similar results be obtained in other boroughs? In inner-city areas? In rural areas?
2. What would be the impact (measured by changes in the teaching of mathematics made by all the teachers in a school) of variations in the timing and duration of the support visits?
3. It would also be valuable to study the effects of having the heads present all the time at off-site working sessions as well as at the on-site working sessions.

Possible surveys of the work of LEA advisers

1. The extent of the provision of mathematics advisers and advisory teachers in LEAs. For how many teachers are specialist advisers responsible at primary level? At secondary level?
2. The number of LEAs which offer scale allowances for mathematics at the primary stage and the total number of schools affected. The range and emphasis of LEA expectations for mathematics co-ordinators and the training given to them. This information would make it possible for good practice to be disseminated.

Epilogue

Does the teaching of mathematics merit special attention?

My own experience of in-service education in mathematics during the past 20 years suggests that such attention is essential if we are to effect lasting improvements throughout the United Kingdom. During the 'project era' of the late-1960s much interest in the projects was generated and vast resources of personnel were deployed, but the effects were not permanent. The enthusiasts obtained other posts (such as wardens of teachers' centres, lecturers at colleges of education). Once the initial in-service education at the teachers' centres (set up for this purpose) was completed, teachers became discouraged as they gradually came to the end of their mathematical resources.

The major problem is that, to date, few teachers have received the kind of mathematical education which would lead them to understand and enjoy mathematics. Most teachers were instructed how to do calculations, and sometimes even how to solve particular types of problems. It is a hard task to be asked to provide children with a mathematical education which is entirely different from their own as children. Teachers need to experience learning mathematics by investigation at their own level before they can become convinced that the method works and is enjoyable. They then need to reconsider the mathematics they teach: the situations and the associated language patterns which must be experienced before the mathematical concepts can be acquired.

To apply these methods in the classroom is not easy unless the teachers are accustomed to working with the children, for some of the time at least, in less formal ways than whole-class teaching. They can best achieve this change with the help of classroom support from the co-ordinator or from an advisory teacher. So their needs are two-fold: to learn by a new method the mathematics they are expected to teach, and to change the techniques of their teaching by providing structured activities to help the children to learn concepts.

If we are serious in our desire to improve the teaching of mathematics we have to face the implications in terms of personnel and expenditure. What is suggested here was substantially endorsed by the findings of the Cockcroft Report (1982).

Reference list

ADELMAN, C., JENKINS, D. and KEMMIS, S. (1976). 'Rethinking case study: notes from the second Cambridge conference', *Cambridge Journal of Education*, 6, 3, 148.

AUSUBEL, D. P., NOVAK, J. D. and HANESIAN, H. (1968). 'Learning by discovery', *Educational Psychology*, 519–28. New York: Holt, Rinehart & Winston.

BENNETT, N. (1976). *Teaching Styles and Pupil Progress*. Somerset: Open Books.

BENNETT, N. (1981). 'Teaching Styles and Pupil Progress, a Re-analysis', *British Journal of Educational Psychology*', 51, 2, 170–86 (June).

COCKCROFT REPORT. GREAT BRITAIN. SECRETARY OF STATE FOR EDUCATION AND SCIENCE (1982). *Mathematics Counts*, paras. 46, 286, 287, 378, 379. London: HMSO.

COHEN, L. and MANION, L. (1980). *Research Methods in Education*. Ch. 9. London: Croom Helm.

ELLIOTT, J. (1978). 'What is action research in schools?' *Journal of Curriculum Studies*, 10, 4, 355–7.

ELLIOTT, J. (1980). 'Implications of classroom research for professional development'. In: HOYLE, E. and MEGARRY, J. (Ed) *The World Yearbook of Education. Development of Teachers*. London: Kogan Page, Chapter 22.

GALTON, M., SIMON, B. and CROLL, P. (1980). *Inside the Primary Classroom: The Nature of Classroom Learning in the Primary School*, pp. 122–6. London: Routledge and Kegan Paul.

GREAT BRITAIN. DEPARTMENT OF EDUCATION AND SCIENCE (1978). *Primary Education in England*. A survey by HM Inspectors of Schools, paras. 5. 64, 8.23. London: HMSO.

GREAT BRITAIN. DEPARTMENT OF EDUCATION AND SCIENCE (1979). *Mathematics 5–11: A Handbook of suggestions.* HMI Series: Matters for Discussion 9, Introduction. London: HMSO.

GREAT BRITAIN. MINISTRY OF EDUCATION (1960). 'Balance of Training', ref. G. 539/517. College Letter No. 14/60, issued by the Ministry of Education, 1st October 1960.

HALSEY, A. H. (Ed). (1972). *EPA Problems and Policies*, 1, 66. London: HMSO.

HARLEN, W. (1977). *Match and Mismatch: Raising Questions.* Report of the Schools Council. Edinburgh: Oliver and Boyd.

HORWITZ, R. A. (1976). An investigation of some long-term psychological effects of open classroom teaching on primary school children in England. Thesis, Yale University.

SUMNER, R. (1975). *Tests of attainment in mathematics in schools.* Monitoring feasibility study. NFER occasional reports, 2nd series, 55–61. Windsor: NFER.

TUPPEN, C. J. (1965). 'The measurement of teachers' attitudes', *Educational Research*, 8, 2, 142–5.